LIFE CYCLES

LIFE CYCLES

REFLECTIONS OF AN
EVOLUTIONARY BIOLOGIST

JOHN TYLER BONNER

PRINCETON UNIVERSITY PRESS

PRINCETON, NEW JERSEY

Copyright © 1993 by Princeton University Press
Published by Princeton University Press, 41 William Street,
Princeton, New Jersey 08540
In the United Kingdom: Princeton University Press,
Chichester, West Sussex
All Rights Reserved

Library of Congress Cataloging-in-Publication Data

Bonner, John Tyler
Life cycles : reflections of an evolutionary biologist /
John Tyler Bonner.
p.   cm.
Includes bibliographical references and index.
ISBN 0-691-03319-6
1. Bonner, John Tyler.   2. Biologists—United States—
Biography.   3. Biology.   I. Title.
QH31.B715A3   1993   574′.092—dc20   [B]   92-41540

This book has been composed in Adobe Galliard

Princeton University Press books are printed
on acid-free paper and meet the guidelines for
permanence and durability of the Committee on
Production Guidelines for Book Longevity
of the Council on Library Resources

Printed in the United States of America

10   9   8   7   6   5   4   3   2

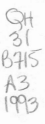

To Ruth

CONTENTS

PREFACE

I AM conscious of an annihilating comment lurking in the wings—that I sound like a professor—but after all those years of professing it is quite impossible for me to totally conceal my principal occupation in life. For over forty years I taught the general biology course for freshmen and sophomores at Princeton University. Once a sociology student came to me and asked me if she could interview me for her work on her senior honors thesis. It turned out that she was comparing teaching with acting. I spent more time quizzing her than vice versa. She made the interesting point that both actors and teachers worry about the response of their audience. However, they have different problems: an actor always gives the same performance but the audience changes every night, while a teacher has the same audience day in, day out, but must say something different (and hopefully riveting) each time. Sometimes the most disconcerting things make a great impression on the audience's mind. I found this out in the very first lecture I gave at Princeton to the underclassmen (at that time there were no women). In the middle of the lecture I slipped off the podium, and fell harmlessly on the floor. There was a great cheer from the students—obviously the high point of the lecture.

Teaching that course for over half of my life gave me an immense amount of pleasure for many reasons, not the least of which were all those students who kept telling me in one way or another what I was doing wrong. As a result the course slowly changed, and at the same time biology changed and I changed. When, not too long ago, I gave my last lecture (feeling rather like Mr. Chips), I decided that since I could no longer teach it, I might "give" the course in the form of a book and in this way prolong the pleasure that had already lasted so long.

That is the way this book began, but it turned out to be the wrong way—another great fall. After numerous starts, I suddenly saw the book I wanted to write. I regret to say it did not come as a shining vision with voices, but rather as a slow dawning. The very scope of that general biology course had made me think about big problems, such as how everything fits together—what is the logic, the meaning of life. In worrying over these grand issues I have had my own evolution, and this book is about that evolution, and more particularly where it has led me: it is my personal life cycle. Inevitably, some of the ideas are ones that I have expressed in previous books over the years, although at various places here I have sneaked in something new. What follows is a resynthesis of my somewhat idiosyncratic views on the fabric of biology, though this time it is not meant only for my biological colleagues, but for everyone else as well.

My readers might well wonder what is new or different about this book. A confession on my part is required: I do not look at biology and its major themes in a conventional manner. I pull biology apart and put it back together in a way that is quite different from that of the average biologist in the street. I do not do this just to play games, or just to be different—I do this because I think many of the more conventional approaches miss the important point as they bear down on details that fit with the views of the moment. My objective is to achieve a more cohesive, more profound view of the science of life—one that has a greater inner consistency and a greater meaning. It is for this reason that the unifying theme of this book is the life cycle.

The appreciation of the central role of the life cycle has been neglected to a considerable degree by biologists over the years. Only recently has there been a new awareness, and this has come from two directions. One is the increasing interest in how genes orchestrate the development of an egg into an adult organism—how they can manage this and what the limits of their power are. The other is the realization that the great discoveries during this century about the mechanism relating to evolution by natural selection must be more than the sorting out of the genes that characterize the adult, for it is not only the gene mixes that evolve, but the entire life cycles of animals and plants. The present and the future of biology involve both the study of genes and adults as well as an understanding of how the mechanics of development connect with the evolution of life cycles.

Since I have made such an effort to avoid turning this modest tome into heavy reading, I was delighted to discover a paragraph in the preface of Mark Twain's *Roughing It*, expressing perfectly how I feel about this book and how I should warn the reader:

> Yes, take it all around, there is quite a good deal of information in the book. I regret this very much; but really it could not be helped: information appears to stew out of me naturally, like the precious ottar of roses out of the otter. Sometimes it has seemed to me that I would give worlds if I could retain my facts; but it cannot be. The more I calk the sources, and the tighter I get, the more I leak wisdom. Therefore I can only claim indulgence at the hands of the reader, not justification.

. . . . .

There are many individuals to whom I became indebted during the course of writing this book, and here I would like to acknowledge their kindnesses and help. First I want to thank Mary Philpott, who helped me teach my course, for suggesting that I undertake this project. The writing was done in a number of different places that I visited during leaves of absence from Princeton. During the early phases I spent a blissful fall semester teaching at Williams College, which, thanks to so many friends there, was a splendid place to write. In the spring of that year (1989–1990), my wife Ruth and I stayed with my brother and sister-in-law in the ancient, tiny annex to their house in a small village near the mountains in Mallorca, Spain. It was not only a wonderful place to get thoughts to flow out of my pen, but I heard nightingales again—a magic childhood memory which I had not exaggerated in my mind over the intervening years. The next winter (1990–91), I was a visiting professor of the Indian Academy of Sciences at the Indian Institute of Science in Bangalore, where I was provided with a splendid office and writing paper that seemed to draw the ink from my pen with even greater avidity. It was there, with the help and encouragement of my good friend Vidya Nanjundiah, that I began to feel I was getting closer to the book I wanted to write. I have finally finished my task here in Canada this summer.

During the course of this one-man-wrestling match, there were many who encouraged me and gave me hints as to how to proceed. From the very beginning Harriet Wasserman had faith in the project, and, in the middle, she had more faith than I had, which sustained me at a low point. Jonathan Weiner came to my rescue at just the right moment for he, both as a former editor and now as a distinguished science writer, understood my problems. He made me see clearly what I was not doing, and gave me the desire and the strength to push on; furthermore, his specific suggestions for improving the text have turned out to be invaluable. Also, my lifelong indebtedness to my colleague Henry Horn continues, as his comments and criticisms were characteristically clear-headed and enormously helpful; for this I give heartfelt thanks. Let me add that a group of anonymous readers, who took my manuscript with them on a field trip, gave me a massive number of useful suggestions that helped to purify my wording and my thoughts, and I thank them one and all. I wish I could say that if anything is wrong with this book it is the fault of all those helpers and friends, but that would not be true; rather, if anything is right in this book, they are responsible.

I dedicate this book to my wife, Ruth. She has a way of encouraging me without my seeing how she does it. This has been true for so many years, and it was no more evident than during my work on this volume. But beyond those undefinable ways she has of giving support, I thank her for all those hours of correcting my drafts. It is a perfect marriage because, unlike me, she can spell and see a mistake in grammar from across the room.

August 1992   J.T.B.
Margaree Harbour
Cape Breton, Nova Scotia

THE BACKGROUND

Chapter 1

BEGINNINGS

I HAVE devoted my life to slime molds. This may seem a peculiar occupation—narrow at best, slightly revolting at its worst—but let me explain why they captivated me and how they opened my eyes so that I wanted to understand not only what made them tick, but how they fit into the general pattern of living things and what the principles are that integrate all of life.

Slime molds are an extremely common organism, widespread all over the world. Yet because they are microscopic and live mostly in the darkness of the soil, they are hard to see, and for that reason they have been little known until recent years. However, if one takes a small bit of topsoil or humus from almost anywhere and brings it into the laboratory, one can easily grow them on small Petri dishes containing transparent agar culture medium. There, through the low powers of a microscope, it is possible to follow their life cycle, which to me has always been a sight of great beauty (fig. 1).

The molds begin as encapsulated spores which split open, and out of each spore emerges a single amoeba. This amoeba immediately begins to feed on the bacteria that are supplied as food, and after about three hours of eating they divide in two. At this rate it does not take long for them to eat all the bacteria on the agar surface—usually about two days. Next comes the magic. After a few hours of starvation, these totally independent cells stream into aggregation centers to form sausage-shaped masses of cells, each of which now acts as an organized multicellular organism. It can crawl towards light, orient in heat gradients, and show an organized unity in various other ways. It looks like a small, translucent slug about a millimeter long (indeed, this migrating mass of amoebae is now commonly called a "slug"). It has clear front and hind ends, and its

Fig. 1 The life cycle of a cellular slime mold. The social phase begins when individual amoebae (*far left*) begin to aggregate into central collection points and gather into migrating slugs. After a period of migration, the slug rights itself and the anterior cells form a stalk that rises up into the air, lifting the posterior cells which turn into spores. (Drawing by Patricia Collins from Bonner, *Scientific American*, June 1969.)

body is sheathed in a very delicate coating of slime which it leaves behind as it moves, looking like a microscopic, collapsed sausage casing.

After a period of migration whose length depends very much on the conditions of the slug's immediate environment, the slug stops, points up into the air, and slowly transforms itself into a fruiting body consisting of a delicately tapered stalk one or more millimeters high, with a terminal globe of spores at its tip. This wonderful metamorphosis is achieved first by the anterior cells of the slug, which will become the stalk cells. They form a small, internal cellulose cylinder that is continuously extended at the tip. As this is occurring, the anterior cells around the top of the newly created cylinder pour into the cylinder, like a fountain flowing in reverse. The result is that the tip of the cylinder (which is the stalk) rises up into the air. As it does, the mass of posterior cells, which are to become the spores, adheres to the rising tip, and in this way the spore mass is lifted upward. During this process each amoeba in the spore mass becomes a spore, imprisoned in a thick-walled, capsule-shaped coat, ready to begin the next generation. The stalk cells inside the thin, tapering cellulose cylinder become large with huge, internal vacuoles; during this process they die, using up their last supplies of energy to build thick cellulose walls. It is a remarkable fact that the anterior cells, on the other hand—the leaders in the crawling slug—die, while the laggard cells in the hind region turn into spores, any one of which can start a new generation. Slime molds seem to support the old army principle of never going out in front—never volunteer for anything.

This entire life cycle (which happens to be asexual) takes about four days in the laboratory. The organisms are very easy to grow, and in many ways ideal for experimental work. The species I have described is only one of about fifty species, making comparative studies possible. Today, in this modern, technical world, one can view one's experiments with extraordinary ease. For instance, I have in my laboratory a video camera on my microscope, and on the screen I can follow the results of any operation I might perform on the migrating slug. If I follow it for two hours, I can immediately play back the changes on time lapse, so the two hours can be speeded up to two minutes. The possibilities make going to the laboratory each day a delight of anticipation. The life of an experimental biologist is one of minute and often humdrum detail involving endless, frustrating experiments that do not work, but the rewards, albeit rare, are great. Suddenly—and how exciting it is when it happens—something will go right and give one a flash of insight into how things work.

A few years ago an old friend who happened to be a veterinarian was sick in the hospital recovering from an operation. While I was visiting him, his surgeon came by and my friend introduced me as "Dr. Bonner." The surgeon asked me, "Are you a small animal or a large animal man?" Without thinking, and somewhat to his alarm, I replied that I was a "teensy-weensy animal man." I have often thought of this episode in the context of the many years I have helped students revive their sick cultures into healthy and thriving ones. One of my main roles in life, then, has been that of a slime mold veterinarian.

. . . . .

My interest in biology began with an interest in birds. My family was living in Europe when I was a child, and I can still remember my excitement at seeing the variety of ducks in St.

James Park in London. (I have to be careful how I put this. Some years ago my university asked me to say something about my first interests in biology for a fund-raising pamphlet. Unfortunately I reported that "it all started watching the ducks in St. James Park." The roommates of one of my sons, who was then a freshman in college, found the quote and immediately taped it to his door.) I was going to boarding school in Switzerland at the time, and with the help of an old pair of binoculars (which my mother had used in her earlier days at the horse races) I began to spend more and more of my leisure time in the woods trying to spot birds and identify them in a rather quaint old bird guide.

My father was quick to see what was happening, and I think he worried at the time that his son, only about eleven or twelve at the time, would find it difficult to make a living as an ornithologist, which I firmly stated was to be my future. Very cleverly he gave me a copy of a wonderful book called *The Science of Life*, written by H. G. Wells, Julian Huxley, and Wells's son, G. P. Wells. H. G. Wells trained as a biologist under Thomas Henry Huxley (Julian's grandfather and the friend and defender of Charles Darwin) at Imperial College in London. Wells gives a splendid account of this in his *Experiment in Autobiography*. T. H. Huxley was the first to preach that there should be a unified biology, not a separate zoology and botany, which was the norm at the time, and for that reason he had an enormous influence on the teaching of "life sciences" in this century. *The Science of Life* is one such unified synthesis that came from this tradition. Wells was keen on bringing everything together under one cover as, for instance, he did in his now forgotten *Outline of History*, and with his two gifted collaborators he wrote a remarkably literate and cohesive biology. The only thing that today seems very odd is a section on mediums and how to communicate with the dead. As with an unpleasant illness, one tends to forget indiscretions of the past, but in the 1930s, when this book was published, many people believed that psychic phenomena of this sort might really exist. Even the great psychologist-philosopher William James left a fund to Harvard University for research on the sixth sense.

Today we look upon séances and mediums, and even extrasensory perception, with either total disbelief or, in the latter case, terminal suspicions, but that apparently was not so when I was a boy. My most vivid memory of this section of the book was a photograph of a woman extruding "ectoplasm" from her nose—large, disgusting masses of it slithering over her shoulder. I think this one picture did much to make a standard biologist out of me; I wanted straightforward, clean animals and plants and stuck to the main part of the book.

My father's plan worked; I soon changed my ornithological aspirations and announced I wanted to become a plain biologist. Reading even part of that huge book was not easy for me, but the more I read, the more captivated I became. I was so inspired that I started to write an illustrated biology book, laboriously typed on a toy typewriter. The most interesting part is the grammar and the spelling (except, of course, for the sections lifted directly from Wells, Huxley, and Wells). At one point my book describes the life cycle of *Paramecium*, the ciliate protozoan, and says it has "two nucleus (*sic*), one for chemical business of every day, the other is responsible for sexual union and other rare reproductive occasions," whatever those might be.

In the years that followed, up through secondary school, I became increasingly involved in collecting everything I could find in the woods and ponds, keeping it alive or carefully preserving it. In this way, without realizing it, I slowly came to have an appreciation of the riches about me, so many of them unseen without the aid of a hand lens or a microscope. It was in no sense an intellectual exercise, but rather the simple enjoyment of poking about in my surroundings.

At the same time, I began reading about various naturalists—the biographies of biologists of the past. I remember being absolutely riveted by the life of Linnaeus, the eighteenth-century naturalist who classified all the known animals and plants, many of them discovered by him. He also devised the system of naming organisms by giving them a genus name followed by a species name, all in Latin (a matter of course at that time)—for instance, we human beings became *Homo sapi-*

ens. Linnaeus lived what seemed to be a splendid life, constantly wandering off from his university in Uppsala, Sweden,
and going on great collecting trips in northern Lapland. How
I wanted to go to Lapland! In fact, I still want to go; the same
old feeling comes over me as I write this sentence! In those
early days I did not delve into the details of his taxonomy—
I was far more interested in him as a person. However, I instinctively felt he held the golden key—and I was quite right.
My reason, however, was not entirely reasonable: it was based
on the fact that he called his great work on the classification of
all animals and plants *Systema Naturae.* Anyone who could put
system into nature was an automatic hero as far as I was concerned. From my puerile collecting I could see that without
some system, nature could seem very chaotic indeed.

It was about that time that Paul de Kruif came out with a
popular book called *The Microbe Hunters.* In it I learned about
Antonie van Leeuwenhoek and Louis Pasteur and found them
tremendously exciting. I was a bit put off by de Kruif's treat-
ment of van Leeuwenhoek; he seemed to be more interested in
his being a church janitor than in the extraordinary things he
discovered with the microscope he invented. Pasteur is still
everyone's hero. He was a man of magic, for whatever he
touched turned into scientific gold. At the same time I read
Sinclair Lewis's *Arrowsmith,* a thinly disguised and highly ro-
manticized account of the early days of biomedical research in
what used to be called the Rockefeller Institute (now University) in New York City. All of this was what I needed in the way
of inspiration to carry me through a lifetime of laboratory research. (I reread *Arrowsmith* a few years ago and could not
imagine what I found so riviting about it in my teens. Certain
books have to be read at just the right age, although I doubt
if a teenager today would be magnetized by Sinclair Lewis's
old book the way I was. It is very dated.)

Perhaps the greatest of my discoveries in the biographies of
scientists was my encounter with Charles Darwin. As is true
for Pasteur, there are many biographies of Darwin and the
drama behind his revolutionary ideas about evolution in the
middle of the nineteenth century. Some of the recent ones are

far superior to the ones I was reading, but nevertheless I was totally captivated. Darwin, like Linnaeus, spent a long time collecting and observing in the wild—not in Lapland, but in South America, Australia, and various islands. This led to his famous *Voyage of the Beagle*, which he published a few years after he returned from his expedition, and a wonderful book it is. He went as a very young man in his twenties and the voyage took five years, but with his extraordinary mind he gathered riches that lasted him a lifetime. He began working on his theory of natural selection a few years after his return, but he did not publish his great book, *On the Origin of Species*, until 1859. He perhaps would not even have done so then, for he was so worried about its reception by the world at large, had not Alfred Wallace sent him a manuscript with the very same idea suggesting natural selection as the mechanism for evolutionary change. Wallace, another gifted naturalist, had the idea while collecting in the Malay Archipelago, and wrote to Darwin about it. Wallace's paper—and a companion one of Darwin's—was published in the *Proceedings of the Linneaen Society*, but the two articles caused little stir. It was Darwin's *book* that created the explosion. Wallace and Darwin became friends, and Wallace was most generous in deferring to Darwin, who had spent so many years working out the evidence for natural selection. As the ultimate expression of generosity, Wallace even wrote a book he called *Darwinism*.

One of the books I read at an early age was Darwin's *Autobiography*, which he wrote primarily for his children. It is an unpretentious account of his life and work in which he emphasized his inability to do well in school and his difficulty in deciding on a suitable vocation. He manages to give the impression that he is an average person with average gifts. Today he is generally regarded as one of the great geniuses of the nineteenth century, although he has always had his detractors who cite, among other things, his autobiography as evidence for his mediocrity. Unknown to me or the rest of the world, the autobiography I read had been greatly censored by his wife, and it was not until the 1950s that his granddaughter, Lady Nora Barlow, published the unexpurgated version. What had been suppressed were Darwin's views on religion: he had, over the course of years, become a nonbeliever. It is particularly fascinating that, in the complete version, he writes that his father, a very successful medical practitioner, was also an agnostic. According to Charles, his father told him that should he ever have doubts, he should not communicate them to his wife for it would make her very unhappy, as then there would be no hope of meeting him again in the next world!

Due to a chronic illness Darwin remained a recluse at his house in Down, Kent. No one really knows what ailed him, but there has been an enormous amount of speculation on the subject. One popular theory is that he contracted Chagas' disease in the Andes. There he was bitten by "the great bug of the Pampas," which through the much later work of Chagas is known to carry the disease. There are also a number of explanations involving his psychological difficulties, and it has even been suggested that he feigned the illness so that he could work without interruption. One result of his isolation was a vast correspondence with others on scientific subjects, only now being edited and published in many volumes. Another is that all the lecturing and debating that necessarily followed the publication of *On the Origin of Species* was done by others, and by far his most eloquent advocate was Thomas Henry Huxley, whom I mentioned a few pages ago. As a teenager I admired Huxley as much as Darwin. In time I came to realize that although Huxley was a superb teacher and an exceptionally gifted essayist as well as a first-rate biologist, Darwin was the one with the really great and profound mind. It is because of the profundity of his thought that his ideas on evolution "by means of natural selection" have such a firm grip on every aspect of biology today and form a principal theme of this book. He provided the intellectual cement to bind all things living that has given us a genuine understanding of the problems of life.

I know I did not appreciate or understand the significance of Darwinism in my youth. That has slowly seeped into the pores of my brain over the span of many years. Observing and

beginning the life of an experimental biologist overshadowed
all else as I entered college and soon found myself in a laboratory. For me it was a heady experience to learn how to weigh
out chemicals, sterilize media in the autoclave, fix and prepare
tissues for microscope observation by imbedding them in par-
affin and slicing them into thin sections on a special machine
with a razor-sharp knife. Some of the techniques I learned do
not even exist today, but to me they were sheer joy. I sometimes have the same feeling of being surrounded by great
riches when I enter a particularly well-appointed stationery
store or a hardware store. I did not do anything remarkable
with these new tools, these new toys; in fact, it never occurred
to me at first to even think about doing the extraordinary. All
I wanted to do was play and then find out if my playing would
allow me to see something I, or perhaps anyone, had never
seen before.

During the second and third year at my university I began
to think of the problem of development—how organisms develop from egg to adult—as one that was not only important
but also interesting to me. This new outlook came from my
courses in animal embryology, where I learned about the ana-
tomical changes that occur during development and was also
introduced to the splendid experiments of Hans Spemann
(and a number of other embryologists, many of them German), who did grafting and cutting experiments to try to understand what Wilhelm Roux called the "mechanics of development." Spemann, for instance, received the Nobel Prize for
showing that one part of an amphibian embryo stimulated or
"induced" another part of it. What he called the "organizer"
region of the early embryo sent out signals to the tissue be-
neath it to begin forming the main axis of the embryo. He
would graft an extra organizer region onto the side of an intact
embryo and it would form two embryos. The importance of
these experiments and the ideas they generated gripped me.
My only difficulty was my considerable incompetence in reading German despite my being half Swiss and, to make matters
worse, all those old classical papers were immensely long. I re-

member bringing a paper home on sexuality in a unicellular alga and asking my Swiss grandmother to help me with it. She took one look at the title and became quite flustered, giving me a long lecture about the general disintegration of young people's moral standards.

While I was dipping into all these embryological riches, I had also decided to do my research work under William H. Weston, a man who had delighted and inspired me from the very beginning of my freshman year. He was not a productive scholar himself, but he was the perfect teacher: his lectures were gems of enthusiasm and humor, and his teaching of research was delicately done by giving his students total freedom, although he was always standing by to help us out of the hole we invariably dug for ourselves. The difficulty with all this was that he was what was then quaintly called a "cryptogamic botanist." That means he was interested in fungi and algae and other "lower" plants, and I knew from the work of his numerous students that he especially favored using them for experimental studies. At first I was torn between animal embryos and lower plants, but it dawned on me that I could combine the two. After all, fungi and algae develop from a single cell, be it a fertilized egg or an asexual spore; why not try to use a simple plant to study the principles of development that had been almost the sole province of animal embryologists?

At first I thought of using a water mold as an experimental "animal," but one day I was sitting in Weston's outer office talking to his pretty secretary when I pulled down one of the many bound Ph.D. theses of his former students and started to thumb through it. It was the thesis of Kenneth Raper on the most extraordinary organisms I had ever seen—ones that nobody had even told me about. They were the cellular slime molds, and Raper had done all sorts of now classic experiments on them. In a flash I knew this was exactly the beast I had been looking for—it was even more suitable than I had imagined could be possible. It was the non-animal embryologist's dream.

In the 1940s these organisms were virtually unknown among biologists except for a handful of cryptogamic botanists. When I first began to give lectures on some of my early work, I had difficulty telling people about my experiments for I was invariably interrupted and deluged with questions about the life history that I described earlier—it seemed to be so unlike that of any familiar organism. I have been told that Arthur Arndt, who made the first motion picture of slime mold development, used to argue in his public lectures in Germany in the 1930s that their life history was so amazing it could only be explained by resorting to vitalism! As a beginning assistant professor, I gave a brief lecture at the Marine Biological Laboratory at Woods Hole in Massachusetts, and apparently word had spread to the world of journalists about this queer organism. Some days later I received a letter from J. J. O'Neill, the science reporter of the *New York Herald Tribune*, saying that he had heard I had done something more important than discovering the atomic bomb: I had created a multicellular organism! I quickly replied that it was not I but God who had managed the remarkable phenomenon and please, for my sake, to restrain his journalistic ardor. My leaning so heavily on support from God may be excused by my anxiety not to be embarrassed by the newspaper (I was not; he wrote a very sensible article). But if I had wanted to give the reason that I really believe explains the peculiar life cycle of slime molds I would have said natural selection, the brainchild of Darwin and Wallace.

I did not need to be convinced that slime molds were the ideal organism for the study of development, for I felt it in my bones. Nevertheless, I received the most tremendous reinforcement in the nicest way. As a graduate student I was invited to Yale to give a seminar. I expected a small group, but to my terror it was a huge joint lecture with both the botanists and the zoologists present. I somehow managed to stagger through it with the help of a film and answers to innumerable questions about the life cycle. Slime molds were doing something unheard of: they grew first and then aggregated into a multicellular organism. After the lecture, what was left of me was gulping a cup of tea when Professor Ross G. Harrison, then in his eighties, came up to me. He was considered by me, and everyone else in the world, the greatest living embryologist—among other things he was the first to grow animal cells in "tissue culture." In his shy and kindly way, he said to me that if he were starting all over again he would work with slime molds. All my strength immediately returned—I am sure I must have glowed in the dark. Even as I write of the incident here many years later, I still feel a warmth come over me.

Chapter 2

THE LIFE CYCLE

WORKING on slime molds at such an early age had an unexpected salutary effect on me. I was very conscious of the fact that my favorite organisms were different, and for that reason alone I should continuously pay attention to other organisms and look for similarities and differences. This led me, with the help of some gifted students, to small skirmishes on other "lower forms" such as colonial algae, fungi of various sorts, and protozoa. Unconsciously I suspect there was something else influencing me. Slime molds had such an obvious life cycle, from germination to the final fruiting body, that I began to realize that development was only a portion of the whole life history of an organism. In slime molds it is a large portion, but in human beings, for instance, it is only a fraction of our entire life span. Slowly something began to trickle into my brain: organisms are not just adults—they are life cycles.

This is far from an original thought. It is one that has been fully appreciated by philosophers for some time. The first time I heard it was in the first lecture of a philosophy course I took as an undergraduate, although it did not sink in at the time. The only thing that did penetrate was the realization that I was not made to become a philosopher. Recently I read a quote from Isaiah Berlin who says exactly what I feel—I only wish I had been able to express it so cleverly: "Philosophy is a wonderful subject, but it is necessarily unfinished and unfinishable. You really can't solve anything. At the end of my life I want to know more than I did at the beginning. And I couldn't get that from philosophy."

It took me a long time to realize that one cannot, through philosophy (as one can with mathematics), find new things in biology. One can only use it to clean up the mess made by philosophically naive biologists. Their use of words, the logic of their sentences, may be hopelessly inconsistent and self-contradictory, yet they may have discovered a new fact or principle that will shine through all the errors in logic and be recognized as a significant contribution. Some years ago a friend who happened to be a distinguished mathematician came to Princeton and we arranged to have lunch. He came to my office a bit early, so I asked him to please have a seat and find some reading matter on the shelves and I would return shortly. When I came back, to my horror he was reading my undergraduate senior thesis, the first half of which is an analysis of the problem of development using symbolic logic. I admonished him for picking it out in a room filled with *good* books. He replied, "Don't be silly, this is wonderful." When I asked how he could say something so absurd, he said, "You don't understand. What's wonderful is that you got this out of your system so early in life!"

The point made in my initial philosophy class concerned the definition of a "dog." (It could have been any other animal or plant, but for some reason dogs are favored for this exercise in precision.) Normally we think of a dog as an object at an instant of time, and that time is usually some moment in its adulthood. Of course we recognize that a young puppy is also a dog, but we begin to waver when we are confronted with a small fetus. Certainly we draw the line if we look at a fertilized egg of a dog. As we peer down the microscope at this undistinguished translucent sphere, nothing in the world would provoke us to say, "Oh, look at the small dog," although that is exactly what it is. The solution to the problem is fairly obvious: if we unstick ourselves from the notion of only using a minute time slice, then we can immediately see that a dog is really a life cycle: from fertilized egg, to maturity when it can begin to reproduce, through old age when it dies of decrepitude. The same is obviously true of human beings; the same is even more obviously true of slime molds. We will continue to think in terms of instants in time because it is very convenient. If we look another person in the eye, we think of that person, that instant, as a human being, and do not wander in our minds through a great course of morphological changes that accompany their progress from egg to adult.

While the important philosophical point may be cumbersome for daily use, it is an essential concept for the biologist who worries about development, evolution, and all the other life processes. We talk often of one generation succeeding another, and what we mean by this is that life consists of a succession of life cycles of organisms. This immediately raises many points of great interest. For instance, why are there cycles at all? Why don't organisms live indefinitely? Why do they have a single cell at one stage in their life cycle, that is, the fertilized egg? What are the consequences of this continued cycling? Let us examine these questions carefully, because they lie at the foundation of all biology.

To begin, it is important to have a straightforward way of categorizing the life cycle and also of visualizing it. There are four main periods in a simple life cycle such as is found in most organisms:

1. There is a single-cell stage. As I have just pointed out, this is usually a fertilized egg, although in asexual life cycles it may be a single-cell spore, as is the case in cellular slime molds. There are exceptions. Some asexual organisms have a multicellular small stage between generations, but such organisms are rare, and as I will explain later they are indeed exceptions that prove the rule.

2. The single-cell stage is followed by a period of growth and development. This is especially dramatic in the case of multicellular organisms where not only does a single cell divide many times, but a shape and an organization emerge, be they in the form of a cow or a maple tree. This period of development is common to all organisms; even unicellular forms have a limited period of development between cell divisions.

3. The period of maturity varies greatly with the kind of animal or plant. In some cases at a certain point, development stops and the adult size is maintained for an extended period of time. This is clearly the case for human beings and many other mammals. However, others, such as the elephant, con-

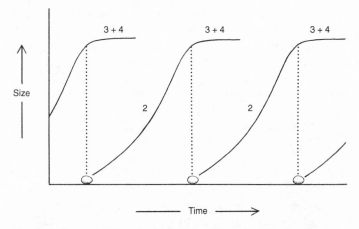

Fig. 2 A diagram illustrating successive life cycles. The fertilized egg develops into an adult, which then produces egg and sperm for the next generation. In this graph the organisms reproduce only once, while in other species there is repeated reproduction. Also, the adult size may continue to increase, as in woody plants, fishes, reptiles, and some mammals such as the elephant.

tinue to increase in size slowly during their years of maturity, and this phenomenon is even more conspicuous in the case of fish and reptiles such as turtles and crocodiles. It is also true for trees, which continue to get larger for many years after they first bear fruit.

4. The period of reproduction generally coincides with the period of maturity. However, this is not invariably the case, with human beings providing a most interesting exception. Our species is unusual in that women have a menopause, giving them an extended period in later life when they can no longer bear children.

For me it is always helpful to visualize concepts when possible. Life cycles can easily be seen as graphs in which the size of the organism is plotted along a period of time (fig. 2). The organism is smallest at the period when it is a single cell (1); it rises rapidly during the period of growth and development (2), and either continues to rise or levels off completely at maturity, depending on the organism (3). The period of reproduc-

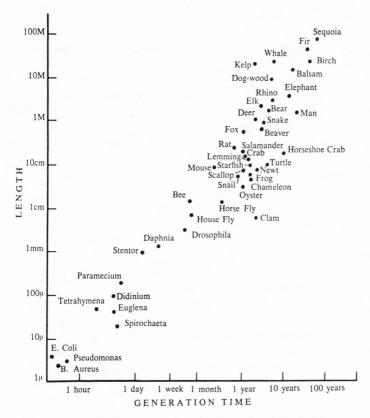

Fig. 3 The length of an organism at the time of reproduction in relation to the generation time, plotted on a logarithmic scale. (From Bonner, *Size and Cycle,* Princeton University Press, 1965.)

tion can be a single event, or a continuous one for variable periods of time (4). Since the time from fertilization to the first production of eggs and sperm is one generation, one can easily see that the smaller the organism, the less time is needed for growth and development. This relation between size and generation time is interesting because it is a very general principle that applies to all animals and plants (fig. 3).

. . . . .

This brings us to the really interesting questions: Why do we have life cycles at all? Why do not all organisms exist more or less indefinitely, like rocks? The answer lies in the natural selection of Darwin, as does all else in biology.

To explain why this is so I must go backward and forward in my explanation. First I will briefly explain the nature of natural selection. In order for selection to occur there must be heritable or genetic variation within a population, and those individuals with a genetic makeup that is most successful in reproduction—in having children and grandchildren—will eventually replace less successful individuals. Natural selection is a selection for organisms in a particular environment who will have the most descendants. This is the forward way of looking at the matter. However, it is much better to look at it backward in time: those variants that exist today have the genes that were favored by natural selection.

I put the matter this way because today we know what Darwin did not know: that genes are the basis of heredity and that genes are stretches of a remarkable molecule called DNA which lies on the chromosomes of all cells. It is the genes that are culled within a population of animals or plants. The matter has been put in its ultimate form by Richard Dawkins in his well-known book, *The Selfish Gene*. Genes compete with one another for survival, and the organisms with the successful genes will pass them down to future generations. Or to put the matter in the backward form: the genes that exist in animals today are the successful ones that have been passed down through previous generations and have not become extinct. All the genes in present-day organisms may not be ancient, however. Genes can undergo mutations and change, or essentially become new genes. These new genes might, in some cases, be well suited for survival in the competition between genes and be passed on successfully to the present day. This means that the animals and plants we see around us have largely ancient genes which survived a long time, but they also have genes that were invented by mutation more recently. The genes in any one organism did not necessarily appear at the same time; there is just a great mixture of genes of different ages. Most of them are successful, the evidence being that they exist today. However, there may be a few genes which have mutated very recently and have not yet been tested to see if they are to be survivors or if they will become extinct.

Let us now return to some mega-questions. Why do we have life cycles at all, or for that matter, why does evolution of living things occur? The answer again is Darwinian natural selection, and I will try to explain why this is so.

One of the most fundamental properties of life is the ability of DNA to make copies of itself by template replication. Without going into the chemical details, I can say that this replication of molecules has, in the origin of life, led to a competition to see which parts of the DNA were most successful not only in making copies of themselves, but copies that survived subsequent replications. There is much more to the problem of the origin of life than this, but replication is a vital element.

If we look at this early molecular evolution backwards, then we see the same thing we saw in the genes of living organisms that live today: some portions of the earlier DNA have survived and others have not. In other words, gene survival or the culling of genes by selection must go back to the very beginning of life, and to do this one must have replication cycles which are in fact life cycles. We have simply made the obvious point that evolution by natural selection could not occur unless there were replication cycles (life cycles) of the DNA.

The more difficult and fundamental point is: Why is there molecular cycling in the first place? If a DNA-type molecule exists (and we will not inquire how—that goes beyond the regression of questions I am asking), then it will, with the help of catalysts, duplicate. The progeny of this replication can in turn replicate, always provided the suitable catalysts are available to make the reaction go, and provided the raw materials are available to make the new chain of DNA. However, there is no infinite supply of the raw materials (the sugars and the nucleic acids) available for replication, and therefore there will be competition among DNA strands for those basic ingredients. The strand that is best at grabbing will be the winner, and among its offspring there may be some mutants that are even more efficient at grabbing the sugars and nucleic acids needed for replication. So already we can see the beginnings of natural selection. As this replication continued, the length of time for a generation must have come under increasing control and regularity, for such would obviously be selectively advantageous. This would be a reasonable scheme for the origin and the regularization of life cycles, which in turn would have been the beginning of evolution.

Again, let us look at the process backward in time. Today we have overwhelming evidence that life cycles, evolution, and natural selection have occurred. Therefore, even if my fairy tales of how they might have started are quite incorrect, we know that they did start, they have existed for a few billion years, and they exist today. We may never know the details of their origin, but the evidence that there *was* an origin is overwhelming, unless one is satisfied with some mystical explanation. In spite of the reasonableness of my arguments, I still have a feeling of awe when I think that if we started off with one origin of life, it has during the last 3.5 billion years radiated out to the millions of different species that exist on earth today, plus all the species that have gone extinct, with their remnants left in the fossil record. The variety of shape, size, and characteristics of all plants and animals is quite unbelievable in its magnitude—did all that start with some simple replication, cycling molecule, or the equivalent? The only solace I find for this disturbing question is that a few billion years is an unimaginably long time.

It has taken me literally my whole scientific life to understand the extent of the implications of natural selection and how it could be the basis of evolution. This may seem odd because the facts that faced me when I was in my teens are essentially the same as the ones we have today. It has taken fifty years for their significances to penetrate.

It is not because I was denied the opportunity. For instance, one summer, when I was a callow assistant professor, Edwin Grant Conklin, the towering American embryologist of the turn of the century and an emeritus professor at Princeton University, heard that I needed laboratory space for the summer at the Marine Biological Laboratory at Woods Hole, and he kindly asked me to share his room, which had been deeded to him for his lifetime as one of the founding fathers of the laboratory. It was a wonderful experience for me because I had not only great respect for Dr. Conklin, but considerable awe. The difficulty was that he loved to talk, and what he had to say was so riveting that it was hard for me to keep up in my own work. But being with him was well worth the price.

At that time Dr. Conklin was in his late eighties, and I realize now that he tried to tell me in his own way that it took him a very long lifetime to appreciate evolution and natural selection even though he had written books on the subject and was known as a champion for Darwinism in America in the early part of this century. One day he came bustling into the room, and in his resonant, preacher's voice he asked me if I had seen the new book in the library on South African fishes which he had just been admiring. He was standing over me as I was sitting by my old brass microscope, and in his excitement his voice became increasingly louder and insistent. He said the book showed the extraordinary VAR-I-E-TY of species of African fishes, and then, pounding his fist on my table so that my microscope jumped, he literally roared at me that "the trouble nowadays is that no one understands the MAG-NI-TUDE of EV-O-LU-TION." By this time I was cringing in my chair. It was only many years later that I began to understand what Dr. Conklin was trying literally to hammer into me. It had also taken him a lifetime to fully appreciate evolution and natural selection, and, bless him, he wanted to start me off in life fully armed.

If it took me a lifetime to ripen my appreciation of evolution and natural selection, as apparently it has taken others, what chance do I have of convincing my younger readers? I do not even have the luxury of pounding my fist on their tables or bellowing into their ears. All I can do is give the facts, pretty much the same ones Conklin and, many years later, I had, and tell you not to be deceived into thinking that because the facts and the theory of natural selection that goes with them are so straightforward, the theory is inadequate to explain the facts.

To return to the life cycle itself, I would now like to examine in more detail each one of its four phases, beginning with the first, the single-cell stage.

The question one immediately asks is: Why do life cycles of a large multicellular organism go to the enormous bother of retaining a minute single-cell stage? It is a complicated and expensive reconstruction job to produce a new adult each generation. Would it not be much easier simply to pinch in two and regenerate the missing half, the way small flatworms (*Planaria*) manage to do so easily? This suggestion always conjures up the image in my mind of a huge elephant pinching in two and at one end remaking a new trunk and tusks, brain, eyes, and so forth, and all the rear parts being manufactured as a bud of the old front half. Disregarding the grotesqueness of this picture, one could answer that it might be easier to start all over again and this is the reason for a single-cell stage. This indeed may be part of the correct answer, but there is a much more important part which has to do with sexuality.

Almost all organisms are sexual, and there has been great interest in recent years as to why. The inspired *New Yorker* writer and cartoonist James Thurber wrote a book called *Is Sex Necessary?* He was more interested in human foibles than in biological evolution, but even at that level the subject seemed interesting and, in Thurber's magic hand, amusing. More recently there have been a number of books which could have used the same title as Thurber's but concentrated solely on the evolutionary problem. Sex is an expensive process, more so than asexual reproduction, because, among other things, it takes two adults to make one fertilized egg. Why then is such a costly procedure so widespread and successful? Here again we are asking a question based from a view backward in time; sex is ubiquitous and therefore the genes for sexual reproduction must have been continuously successful over millions, perhaps billions, of years. The answers of contemporary scholars are varied as to what might be the selective advantage of clinging to a kind of reproduction that is not only elaborate and often cumbersome, but invariably expensive (compared to asexual reproduction). There are a number of possible explanations that have been offered for this paradox, all of which may be true. There is one, however, that perhaps is the most significant of all, one that has been appreciated for some time. It is that sexuality is enormously effective in seeing to it that the amount of variation produced in offspring—that is, from one generation to the next—is neither too much nor too little but optimal for evolutionary change. It is a system which, under the eagle eye of natural selection, has been preserved for eons of time. Just the way a regimented cycling was favored by selection in the beginning of life histories, so was sexuality favored. That is, the genes which governed sexuality were always successful ones, for through their actions it was possible to produce organisms that survived and had many descendants.

Not only has sex been greatly successful through natural selection, but the particular system of handling genes on chromosomes, of duplicating and recombining them—that is, their controlled mixing or shuffling during the process of meiosis and fertilization—has essentially been the same since the time it was invented in primitive organisms. One should think of the system as an intricate but extraordinarily effective way of handling and controlling variation—so effective that it has not only been continuously selected for during a large share of the time of life on earth, but it has remained in essentially its original form. The essence of shuffling is found in the cell processes of meiosis and fertilization, both of which, from a mechanical and biochemical point of view, are exceedingly complicated. If we look at simple and primitive organisms found on earth today, along with the middle-sized and large, most complex ones, we see that all have essentially the same basic mechanisms. Therefore, not only has sex itself been successful, but the specific, detailed mechanism of how it is carried out, once invented, has been kept through evolutionary time, for virtually all organisms. From the point of view of natural selection, this is a stunning success story.

Let us now turn to the exception that proves the rule, that is, asexual reproduction. Many simple algae, a number of invertebrates, and a number of higher plants have both sexual and asexual reproduction. A good example is an alga such as

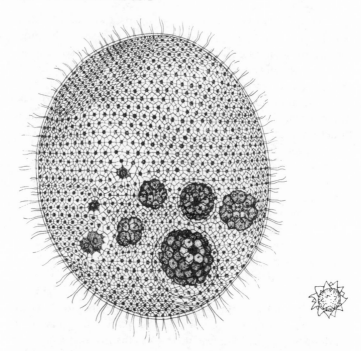

Fig. 4 *Left:* A colony of *Volvox,* showing the successive stages in asexual development of a daughter colony. *Right:* A fertilized egg (much magnified) showing its thick, resistant cell wall.

Volvox, which consists of beautiful green spheres of cells that are found in pond water (first described by van Leeuwenhoek in the latter part of the seventeenth century) (fig. 4). In the spring and summer *Volvox* reproduces asexually; certain cells in a colony begin to divide and form daughter spheres within the mother colony (as van Leeuwenhoek observed). The advantage of this inexpensive kind of propagation is that many generations can be produced in no time at all: the sunlight is plentiful for photosynthesis, and the ponds are warm, favoring quick growth. These conditions give rise to the algal blooms one finds in some ponds during the warmer months. In this case asexual reproduction seems to be taking advantage of the good weather and the relatively constant conditions to produce as many offspring as quickly as possible. At the end of the summer season, when the weather begins to turn, *Volvox* changes its development and opts for sexual reproduction, producing egg and sperm (from different colonies) that produce a fertilized egg. This cell does not turn into a new colony—it undergoes a series of changes, producing finally a resistant, dormant stage in which the cells are encased in a thick cellulose case, often of a peculiar, spiky shape (fig. 4). This resistant stage in its cellulose armor can now survive the rigors of winter as it lies deep on the mud bottom of the pond. It does not awaken, like everything else, until the spring, when it swells and the cramped cells expand to produce a new colony to bloom by asexual reproduction in the summer.

Note that in the asexual stages there is little or no genetic variation. Unless there is a chance mutation, identical clones are formed from one parent colony. However, in the fall the resistant product of sexual reproduction has the benefit of recombination of genes from two parents, and this variation could conceivably be of significance to meet the challenge of the new warm months to come. The strategy is that if you have a good gene combination and the conditions for growth are ideal and constant, then multiply as rapidly as possible without the expense and complications of gene shuffling. As soon as there is any uncertainty as to the future, hedge your genetic bets for the next season by producing genetically variable offspring.

This general strategy of getting the advantages of both sexual and asexual reproduction is by no means confined to *Volvox*: it is a quite general phenomenon. For instance, aphids, water fleas, and many other arthropods (i.e., insects and crustacea) can produce eggs that are unfertilized, a process known as parthenogenesis, which is a form of asexual reproduction. Again, they reproduce this way when conditions seem to be constant and predictable; as soon as the uncertainties of fall envelop them, they mate and produce fertilized eggs. These sexual eggs often have a thicker coat and are more resistant to the devastations of winter. The phenomenon is also common among higher plants, although they grow so slowly that asexual development is not a seasonal phenomenon but can extend over many years. Plants such as aspen, huckleberry, bamboo, and many others propagate through runners so that one original plant can produce a clone that not only may be very ancient, but covers a large area of ground. The species that do this can usually also reproduce sexually with pollen and ultimately produce a seed that contains a fertilized egg. Therefore, they do have a system of producing genetic variation when new conditions require a change for survival.

Let me add that the kind of slime molds I have talked about so far also have a sexual cycle. (There is another kind that is quite different. Technically these molds are called myxomycetes, while the slime molds I work with are called cellular slime molds. For convenience I have been and will continue to use the term "slime molds" for cellular slime molds, and I will call the others myxomycetes.) The sexual cycle has been discovered only recently. Normally the life cycles are asexual, but under certain environmental conditions, and if strains of opposite mating types are present, cell fusion will occur, producing the equivalent of a fertilized egg which then undergoes meiosis before becoming encapsulated in a thick, resistant coat. Presumably slime molds employ the same strategy as *Volvox* and many other lower forms by having a great succession of asexual cycles, interspersed with occasional sexual ones. (Here and in later pages I use the words "lower" and "higher" simply to denote less and more complex animals and plants, without any other implications.)

Organisms which have lost their sexual cycle completely, such as quite a few algae and fungi, will probably be safe in the short run; but if there is a major shift in climate, they will be the first to go extinct because their ability to handle variation effectively has been lost. Presumably they have existed in a constant environment for so long that they have neglected to keep their sexual cycle, which is so essential for their long-term existence.

In asexual cycles the point of minimum size is often a single cell. In fungi and slime molds these cells exist in the form of unicellular spores, which no doubt are small to facilitate dispersal, a phenomenon that is clearly of great significance for their long-term survival.

There are, however, instances where the point of minimum size in the life cycle is made up of a structure that, although it is small, is nevertheless multicellular. An interesting example can be found among lichens, the flaky crusts one sees on trees and rocks, often in unfriendly climates such as the far north. They consist of two organisms living together. Their external structure is that of a filamentous fungus, and contained within this fungus house are many cells of an alga. There is reason to believe that the partners of this dual organism benefit each other: the alga produces carbohydrates by photosynthesis, and the fungus produces a structure that supports and protects the alga. Now, if such a dual organism is to disperse, it must have a "spore" that contains both partners. This is exactly what one finds: they cut off a small package of cells made up of both the fungus and the alga that is liberated, and by wind or other agents it is carried to new locations which it can colonize. This is indeed another exception that proves the rule: since it involves an organism made up of two separate species, both species must be in the small structure that is dispersed to start the next generation.

To turn next to the second phase of the cell cycle, the phase of size increase or development, this is such an important subject that I will devote an entire chapter to it (chapter 4). Here I shall just point out that there are many ways by which the size increase of development can be achieved, and this variety in the modes of construction is one of the reasons the subject is so fascinating. The best way to understand why there is such a variety—which at the same time will give us insight into the problem of development—is to consider the different origins of multicellularity (the subject of chapter 3).

The period of maturity in the life cycle is the period when animals and most plants do certain things, such as respond to the environment by orientation towards or away from chemical substances or light or a source of heat (chapter 5). Animals in particular show a wonderful sensitivity to external stimuli, which goes hand in hand with a nervous system and brain and ultimately produces an organism capable of complex behavior (chapters 6, 7, and 8).

There are many other activities that are carried on during the period of maturity which we will not consider in any detail. An adult organism functions in many ways. It not only responds to stimuli, but it can move and it can process energy to make all the body functions possible. Animals are especially active: they breathe to take in oxygen and eat to take in and capture the chemical energy which lies in food. They can channel that energy into muscle for locomotion, which allows them to catch food or escape predators. The study of these functions of living organisms is called physiology. Each function has evolved by natural selection and serves to increase the chances that the individual can reproduce successfully in a very competitive environment. While physiology is a subject of great interest and importance, it is not one I will cover exhaustively. This is partly because the topic is too vast for any comfortable discussion, and partly because I want to stay close to the evolutionary consideration of life cycles.

One important aspect of the period of maturity is that we are all, young and old, fascinated by the subject of age and overlapping generations. I can remember my Swiss grandfather telling me that he remembered hearing Brahms conducting the Zurich orchestra. At my young age I found this almost impossible to believe—it was like someone telling me he could remember seeing dinosaurs. Yet I have a clear recollection as a boy watching Memorial Day parades and seeing considerable numbers of Civil War veterans, all done up in their uniforms, marching along with the veterans of the First World War—my father's generation. My only disappointment in this reaching back into the past came when I was in my early teens and went to a country fair. For twenty-five cents one could "shake the hand of a man who had shaken the hand of Napoleon." This seemed well worth the large investment of a quarter. But, alas, when I got close to this grand old man I could see that all his beautiful wrinkles had been painted in, and I realized immediately that I had been taken. Such are the things in life which hasten maturity.

There are some organisms that never seem to age. This is true, for instance, of sea anemones, which are related to jelly fish and freshwater hydra. The world's record for longevity is a group that lived in the Zoology Department at the University of Edinburgh for somewhere between eighty and ninety years. Even at their advanced age they appeared no different from a young sea anemone. I have always had a special feeling for these beasts because of their name. When I first started in biology, all my knowledge came from reading, and for years I pronounced their name incorrectly as "sea ani-moans." This still seems to me a more reasonable pronunciation.

Many animals never have time to show any signs of aging. Their environment is so competitive and harsh that they are killed while still young and never exhibit decrepitude. Small birds, for instance, such as sparrows and warblers, or small mammals such as fieldmice or shrews, on the average will live no more than a year, just enough time to produce one set of young in the spring. However, if they are carefully cared for in captivity, they live much longer and, unlike sea anemones, do show signs of senescence. It is sometimes argued that trees, since they continue to grow indefinitely in height and in girth, do not senesce. This is clearly not the case. A well-cared-for apple tree, for instance, will produce an increasingly large crop of apples for many years, but then the crop will start to decline and the tree shows signs of creeping decrepitude. One could argue that this is due to a size increase and a changing weight-strength ratio, making the tree more susceptible to wind, icing, and the various perils of a fluctuating environment. No doubt there is some truth to this, but the end result is one of obvious decline. Furthermore, we know that each species of tree has a maximum size—giant sequoias are immense compared to apple trees.

Senescence is very obvious in human beings, and in fact in all domestic animals. By simply glancing at a person we can estimate his or her age reasonably within five years. The signs of wear and tear in later years are just as obvious to the eye as are the features of earlier growth and maturing. They are as evident in old dogs and cats and horses—each in its own way shows signs that its body is wearing down. These examples are at the opposite end of the scale from sea anemones; in fact, sea anemones are the rare exception. Signs of decay are the rule if chance preserves an individual from an early accidental death.

The great German biologist August Weismann suggested in the nineteenth century that senescence had arisen by natural selection: there is a selection pressure for a limit on the life span, so that genes which contribute to decline are selectively advantageous. In this way parents and grandparents will not be competing with their descendants for limited resources, a condition that would impede the chances of continuous reproductive success in succeeding generations. Unfortunately, Weismann phrased the matter in such a way that Peter Medawar, in a brilliant essay, accused him of making an argument in which "he canters twice around a vicious circle." Consequently, Medawar and a number of others argued that selection does not occur directly on limiting the life cycles or the age of maturity on the cycle, but that senescence is an indirect consequence of selection. They suggested that some mutant genes may have appeared which were advantageous to the growing organism during its prereproductive phase, and that these same genes have a deleterious effect later in the life span. By the accumulation of such genes with good early effects and bad effects later in the life cycle, senescence would result.

Another equally convincing argument is that if any harmful gene mutations appear late in life, they will not be selected against because they occur so late, after the period of reproduction, and therefore will not be passed on to the next generation. If they appeared before the reproductive period, they would be rapidly eliminated because those individuals would fail to reproduce. This is presumed to be the explanation why in human beings devastating cancers are more likely to appear after an individual has reached the age of forty, after the period during which most childbearing is likely to have occurred. The same is true for Huntington's chorea, a horrible genetic disease which causes the degeneration of the nervous system and death.

If we again look at the issue of senescence backward in time we see, as did Weismann, that all organisms that live today have a limited period during which they can reproduce, and once reproduction is accomplished these adults disappear. This may happen because the harsh environment they live in makes prolonged survival unlikely (as in small birds or mammals) or because they have accumulated genes which have a deleterious, life-shortening effect in the late reproductive or post-reproductive period of their life cycle. There are extreme cases, such as in the mayfly or the Pacific salmon, which reproduce only once, right after spawning. This death clearly must be genetically determined; the adult is no longer needed for the propagation and guarding of the genes for the next and subsequent generations. Therefore any mutation that might cause death could filter in; there is nothing in the form of natural selection that would prevent it.

I would briefly like to discuss one final aspect of the period of maturity in the life cycle. Most organisms, but by no means all, are active in one way or another during adulthood. This is quite obvious in animals which rush about to find food, to mate, and to escape predators, or which migrate to escape inclement environments. It is also clear that sedentary plants such as trees are quite active in photosynthesis and respiration even though they perform these metabolic activities in a totally immobile state. There is, however, a large group of annual plants in which maturity means death for most of the plant, and suspended animation for the rest. Consider, for instance, any grass or cereal, such as wheat: it will grow to a meter or so in height and bear seed at its tip. The seeds are a dormant stage that can last through the winter (or remain viable for many years if they are carefully stored), while the stalks that hold the seeds up into the air will turn beautifully golden, reflecting the fact that all the cells of the stalk are dead. Another example of the same phenomenon is the fruiting bodies of slime molds, or the fruiting bodies of many other small molds, and of large fungi such as mushrooms. There is a radical difference in the two kinds of periods of maturity: in one there is activity, when

animals often bustle about; in the other there is total quies-
cence by death and dormancy. Animals compete actively as
adults, while slime molds and many plants compete passively,
concerning themselves solely with spreading their spores or
seeds as effectively as possible each generation. Both of these
competition strategies are clearly successful, for each exists in
such abundance today. Furthermore, by being so different
they occupy and exploit separate niches, enabling them to co-
exist peacefully. Indeed, they often help one another: one provides fodder and the other provides help in dispersal and fer-
tilizer for more vigorous growth. Through time they have
become utterly dependent upon one another.

Finally, we return to the fourth stage of the life cycle,
namely reproduction. It is the link from one life cycle to the
next; it is the prime basis of the cycling itself. I have already
emphasized that most reproduction is sexual and that asexual
reproduction is either an alternative form of reproduction, as
we saw for *Volvox* and slime molds, or all the reproduction is
asexual because the sexual reproduction has been lost. In other
words, sexual reproduction came first; asexual reproduction
was invented later to produce fast, inexpensive reproduction
when the conditions are constant and favorable.

Since sexuality is the prime method of reproduction for all
organisms, there must be a single-cell stage for all life cycles.
(Bacteria and related organisms do things slightly differently,
as I will discuss in the next chapter.) The reason for this is
quite obvious: to bring the genes of both parents together,
there must be one set of chromosomes with genes from the
mother and another set from the father; they fuse to give the
offspring its unique combination of genes from both parents.
This can only be accomplished by two cells, the egg and the
sperm, fusing to form the single initial cell of the next generation, the fertilized egg.

This initial cell will now begin to develop by a series of cell
divisions (growth) and by the formation of all sorts of cell and
tissue types (differentiation) which will be apportioned in an
appropriate shape or pattern. One question that immediately
arises is: Why do organisms become multicellular—why is the

earth not simply populated with the ancestral unicellular forms? The answer must be that in the competition for successful ways to reproduce, size increase has led to certain advantages that have resulted in a strong selection pressure for larger size. Larger organisms, compared to smaller ones, occupy different niches and in this way can be assured of successful reproduction. One could argue then that development is the inevitable result of sex and size. The single-cell stage is required for sexual reproduction, and the larger size is the result of selection for reproductive success in new niches.

THE PERIOD OF SIZE INCREASE

Organisms are life cycles. They arose as a result of natural selection, and it is natural selection that caused them to change, that is, to evolve. This is the way the great variety of plants and animals on the surface of the earth came into being. One of the giant steps in that enormous span of evolution was the origin of multicellularity, which we will now examine in some detail.

Chapter 3

BECOMING LARGER BY
BECOMING MULTICELLULAR

THE conventional wisdom has always been that there was one origin of multicellular animals that gave rise to the invertebrates and another that gave rise to plants. This hypothesis might even be true, but it ignores a host of primitive multicellular organisms that certainly did not arise from one multicellular ancestor. Again the obviousness of this point thrust itself on my mind because I could not argue, even with the wildest stretch of the imagination, that slime molds are in any way ancestral to higher animals or plants.

When I give lectures people often ask me, "Are slime molds animals or plants?" My answer has to be "neither"; they are simply lower organisms. They have some characteristics of animals (an amoeba-like cell) and some of plants (cellulose walls of the stalk), but this does not put them halfway between. I remember years ago, when there was a big rivalry between botanists and zoologists, a distinguished botanist urged me to claim them firmly as plants, otherwise the enemy might steal them. The senselessness of these arguments is underlined by the fact that traditionally slime molds were found in botany texts and not in zoology ones. The reason was wonderfully absurd. Originally, in the nineteenth century, slime molds were discovered and described by botanists, because their fruiting bodies looked so much like those of bread mold, which is a fungus (and a certified plant with rigid cell walls). So slime molds came into the province of botanists because they found them first—rather like putting one's national flag on a newly discovered island. Human nature seeps into the strangest places. (When I was a graduate student I discovered a beautiful group of unfamiliar fruiting bodies on the door of one of the stalls in the men's room. I was thrilled and immediately asked my professor to come and admire them. He gently informed me that they were not slime molds, but the egg cases of a lovely insect called a lacewing. If I had not been stopped in time I might even have tried to include lacewings among the plants!)

I will now take my reader on a tour (with pictures) of the different kinds of simple, multicellular lower organisms. It will be a short course in natural history of the organisms of which many usually go unmentioned in a biology course. There will be no lions, tigers, and giant sequoias—only the shady underworld of living creatures.

Each example will be of a different kind of organism that has become multicellular to form a simple colony. This has an important implication: that there has been a selection pressure for size increase. Now and later I shall make suggestions as to why size increase might be an advantage in the struggle for reproductive success; but let us momentarily take it for granted that it is true. It is the same assumption I made to account for the increase in size during the period of development that I described in the last chapter.

However, it is important always to keep one point in mind. While the world is populated with many different kinds of primitive colonial multicellular organisms, the single-cell descendants of their ancestors also exist today and are abundant. There has not just been selection for size increase, but a continuous selection for small size as well. The reason is that different-sized organisms occupy separate ecological niches. We will return to this point when we discuss very large organisms in a later chapter.

Let me begin the discussion by pointing out that multicellularity began in two quite different ways. In one a cell divides and the two daughter cells remain stuck together. As they in turn divide and remain in contact, the colony becomes larger. We can call this simply size increase by *growth*. The other method of becoming multicellular is by *aggregation*, of which slime molds are a prime example. But by no means are they the only ones—there are others of great interest.

The next general point of importance is that among lower organisms there are two quite distinct cell types: bacteria-like cells called "prokaryotic" cells, and cells like the ones in our own bodies, called "eukaryotic" cells. Prokaryotic cells have no nucleus, and the DNA is a circular strand that simply lies bunched up in one spot in the cytoplasm. Eukaryotic cells, on the other hand, have a nucleus bounded by a membrane, and the DNA is attached to different proteins that bind it into groups of DNA called chromosomes. There are other differences as well, but clearly the bacterial, prokaryotic cell is the more primitive and, therefore, the ancestral type. There are two reasons for saying this. One is that the very earliest fossils are evidently prokaryotic and are in the order of three and a half billion years old; on the other hand, the first evidence for eukaryotic cells is more recent—about two billion years old. The other reason is that the structure of a eukaryotic cell is far more complex than that of a prokaryotic one. Not only does it have a nucleus, but it has many other complex internal bodies as well. One of them is a mitochondrion that processes the energy in a eukaryotic cell, and there are good reasons to believe that it is the descendant of ancient bacteria that inhabited the insides of eukaryotic cells (like the symbiosis between an alga and a fungus in lichens), and that over eons of time it has lost some but not all of its bacterial characteristics. For instance, it still retains a ring of DNA closely resembling bacterial DNA.

Before I begin my bestiary of these primitive forms, I would like to point out that even with the small increases in size found among the examples we will see, there are a number of cases where we find a division of labor among the cells: not all cells have the same function. Division of labor is intimately associated with size, a subject that I find extraordinarily interesting. It is a principle that applies not only to odd microcolonial organisms, but to all organisms, and extends into the behavior of higher organisms including ourselves. It even applies to the institutions we build: the larger the corporation, the greater the number of individuals with specialized tasks; the larger the village, the greater the division of labor in the form of cobblers, bakers, tailors, butchers, doctors, blacksmiths (or automobile mechanics), electricians, plumbers, carpenters, and so forth. In a small village people can and pretty much have to do everything for themselves. In large animals and plants there is a great division of labor among the cells: we have nerve cells, muscle cells, gland cells of many sorts, lens cells, skin cells, and many more. Each cell type performs a different function—but more on this fascinating theme later. Here I want only to make clear that primitive colonial organisms are gripped by the same principle.

First let us consider those multicellular forms that arose by growth. There are a number of examples among bacteria, which do little more than form flat plates of cells stuck together to form a sheet. But bacteria are not the only prokaryotes: there are also the photosynthetic cyanobacteria, or what used to be called the blue-green algae. They differ from ordinary bacteria in that their cells are much larger, yet the internal structure of their cells is clearly prokaryotic, including the absence of a true nucleus, with a bunched-up ring of DNA in its stead. When these larger cells divide, the products of the division do not separate but are rigidly cemented together. They, like ordinary bacteria, can form colonies in flat sheets, or even in rather perfect cubes, but these are the rare exceptions. Most cyanobacteria form colonies that are straight filaments in which cell division is always in a line producing a single row of cells (fig. 5).

The remarkable thing about cyanobacteria is their incredible toughness. They can persist, and even grow, in the most inhospitable environments and manage quite happily where nothing else can survive. I remember being shocked to discover that one of my teachers, who was an authority on cyanobacteria, would take a tuft of their filaments and allow them to dry on a herbarium card. In this form they could be kept alive. All one needed to do was to scrape a bit off, add water, and immediately (if one had not kept them dry too long) they looked normal again and were capable of further growth. This could hardly be done with any other organism unless one collected the seeds of a plant, or during the dormant stage of some animal.

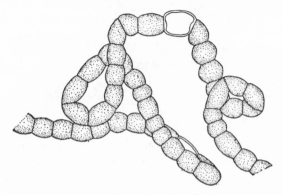

Fig. 5 A filament of a species of cyanobacteria shows a division of labor between the cells in which photosynthesis takes place and the clear cells that specialize in fixing nitrogen from the atmosphere into nitrogen compounds that are needed for growth and respiration. (From J. E. Tilden, *The Algae and Their Relations,* University of Minnesota Press, 1935.)

The ability of cyanobacteria to survive extreme conditions can be seen in nature: they are found both in polar waters and in desert pools where virtually nothing else can grow. They are also known to be very ancient, and it is thought that these survival properties are the very features which allowed them to exist in the severe environmental conditions of the earth of billions of years ago. When the volcano on the island of Krakatao erupted in 1883, not one organism survived—all life had either been blown away or cooked in the intense volcanic heat. As a result, biologists began to watch its recovery carefully. As the barren island cooled down, they discovered that the very first organisms to reappear were the rugged cyanobacteria. They have not lost their skill in invading even the most unfriendly environments.

Cyanobacteria can also divide labor. Some of them occasionally have thick-walled cells which appear to function as spores and shepherd the organisms through especially bad periods. They also have heterocysts, remarkable cells that look rather clear and void of chlorophyll on the inside, but have a very special function. Like all organisms, cyanobacteria need nitrogen to make essential proteins and nucleic acids (for the DNA and RNA) as they grow. Many organisms can get nitrogen from compounds in the soil or in the food they eat, as in the case of animals. But cyanobacteria have to catch the nitrogen gas from the air and make nitrogen compounds with it. This so-called nitrogen fixation involves a chemical reaction that must take place in the absence of oxygen. Since the cells of cyanobacteria are loaded with chlorophyll and convert the carbon dioxide of the air into sugars and oxygen, photosynthesis cannot take place at the same time as nitrogen fixation. In more primitive cyanobacteria—and certainly this would be true of ancestral unicellular forms—nitrogen fixation can take place only at night when it is dark and no oxygen is present, because photosynthesis is impossible without light. However, in the more advanced, filamentous forms, most of the cells make sugars by photosynthesis; but there are a few cells, the heterocysts, that are incapable of photosynthesis and have become specialized nitrogen-fixing cells (fig. 5). Here is a perfect case of division of labor where two functions that cannot occur in the same cell can now occur in specialized cells which lie side by side. It is like going from a small community, where an individual will double as baker and blacksmith, to a larger one, where separate individuals perform these quite different tasks.

A different experiment in multicellularity may be seen in the green alga, *Hydrodictyon*. It has such an unusual life cycle that it is a perfect example for showing that multicellularity arose in many different ways. Nothing except the close relatives of *Hydrodictyon* even remotely becomes multicellular in a similar fashion.

Hydrodictyon, as is true of all the examples I will give for the separate origins of multicellularity by growth, is a eukaryote. It has both a proper nucleus and mitochondria, and the chlorophyll for photosynthesis is packaged in special bodies called chloroplasts, which are thought to be descendants of photosynthetic cyanobacteria that began as symbionts within the cytoplasm, just as mitochondria are descendants of nonphotosynthetic bacteria. *Hydrodictyon* lives in freshwater ponds, lakes, and slow-moving streams. It can become quite large but its shape is always the same: a sausage shape made up of a lace-like meshwork of elongated green cells, hence its common name "water net" (fig. 6).

Each cell grows more or less continuously if the conditions are favorable for producing a larger and larger net. The cells start off as uninucleate, but as they grow the nucleus keeps dividing. However, no cell walls are formed around the nuclei, so that ultimately a large multinucleate sausage-shaped cell is formed, pushing its ends against the ends of similar cells which make up the meshes of the net. When reproduction occurs, each nucleus sprouts a cell membrane that encloses some cytoplasm about it, as well as two whiplike flagella. These cells begin to swim about inside the vacuole of the large parental cell. After their gyrations, they plaster themselves against the inside wall of their mother cell and their flagella disappear. They soon start growth by pushing against the ex-swarmer cells they touch, and each one begins to show signs of a sausage-shaped daughter cell. As they grow, their nuclei go through repeated divisions. This new growth is too much for the old mother cell wall, which bursts open to liberate a new *Hydrodictyon* colony. The newborn colony is sausage-shaped because that is the shape of the mother cell which serves as a mold for the swarmers of the newly forming daughter colony. I have always wished that I could be clever enough to introduce the swarmers into a small cubic box—and have the colony emerge as a cubic net.

It is plain that *Hydrodictyon* and its smaller relatives have devised a very unusual, not to say peculiar, way of becoming multicellular. It is unlike any other organisms, so much so that it must have been invented independently; it is impossible to imagine anything else.

We have already discussed *Volvox*, a quite different kind of green alga. It develops in a much more conventional fashion: one cell keeps dividing to produce a sphere of uninucleate daughter cells. The interesting thing about *Volvox* is that today there are simpler forms—much smaller colonies consisting of far fewer cells (fig. 7). In these small species any cell in the colony can become reproductive and start a new generation.

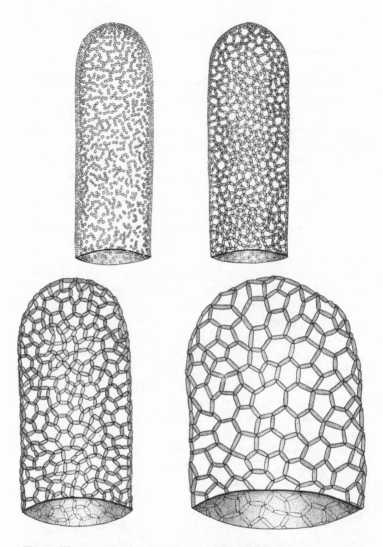

Fig. 6 The growth of a new colony of *Hydrodictyon*. At the upper left the swarmers have settled down on the inside of the mother cell wall, and the growth and elongation of those original swarmer cells produce a new colony which soon bursts free of the old wall. (From Bonner, *The Evolution of Development*, Cambridge University Press, 1958.)

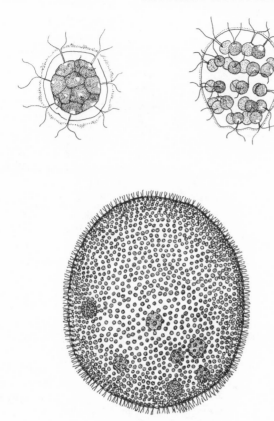

Fig. 7 Three different-sized colonies of Volvocales. Upper left: *Pandorina*. Upper right: *Eudorina*. Below: *Volvox*. (From Bonner, *On Development*, Harvard University Press, 1974.)

This, however, is not true for *Volvox*; only a few special, large cells can serve this reproductive function. In other words, the larger *Volvox* has a division of labor among the cells which is absent in the smaller relatives. A most interesting mutant of *Volvox* has been discovered recently in which all the cells can reproduce, and this mutation involves the change of only a few genes. Perhaps we are glimpsing here at the very first step of the invention of a division of labor for these organisms.

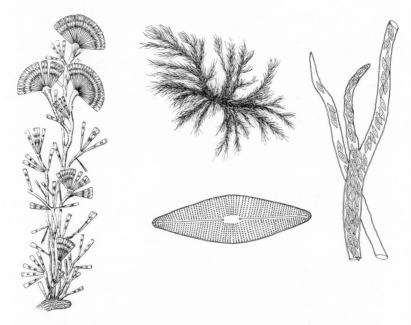

Fig. 8 Two different kinds of colonial diatoms. (A) A species that has a
gelatinous stalk and the daughter cells do not separate. (From Oltmanns,
Morphologie und Biologie der Algen, Gustav Fischer, 1904.) (B) *Upper left:*
An entire colony of another species (the colony is about 1 cm long). *Below:*
A higher-power view of the diatoms wandering about inside the tube they
secrete and enlarge, and a high-power view of an individual diatom.
(Drawing by M. La Farge.)

Diatoms are prevalent in fresh water and in the oceans. In
fact, most of the photosynthesis in the world does not take
place on land—something we would naturally assume since we
are surrounded by grass, shrubs, and trees—but in the ocean
by unicellular diatoms. They are beautiful organisms enclosed
in elegant, sculptured silica shells of an enormous variety of
shapes. Almost all the species are free-living except for a few
forms which are colonial. The more common colonial diatom
is one that is attached to the floor of the pond or ocean by an
adhesive stalk. When the cell divides, the daughter cells remain
attached to a common stalk, sometimes resembling an elegant,

miniature oriental fan (fig. 8A). In another species one finds something far more out of the ordinary. The motile diatom cells secrete a hollow cylinder around themselves, and as the diatoms within this branched tube multiply by cell division, the plantlike tube expands, looking finally like a small seaweed (fig. 8B). If one watches the tube with a microscope one can see the cells scooting about within the house they have built. This is quite an extraordinary step towards multicellularity— and a mysterious one as well. We have no idea how the tubes are built and shaped, nor any clue as to what might be the advantage to the individual cells. One thing is clear: diatoms have such a specialized construction that their multicellular colonies could only have been isolated inventions of their own.

Ciliate protozoa are extremely common organisms, and no one who has taken elementary biology has escaped a chance to admire *Paramecium.* Everyone who has looked at a bit of pond or swamp water in grade school has seen ciliates buzzing about under the microscope. They are peculiar, large cells with a correspondingly large nucleus, and their outside surface is covered with many small flagella (more properly, cilia) which help them to zoom about with great apparent speed.

In common with diatoms they have very specialized cells and they also have stalked forms which, instead of swimming free, use their cilia to waft the food into their gullet. In some of these species, as in diatoms, the daughter cells do not separate but remain attached to a common stalk. In some species there is a common muscle thread or fiber which passes down into the stalk and joins all the cells (fig. 9). If one touches one of the cells with a needle, a wave of contraction will pass through the entire colony so that it will contract to a small cluster around its attached base. Presumably this is a means of avoiding or confounding predators. Again this is such a totally different kind of multicellularity that it can only have arisen independently.

There are other examples of the unique formation of colonies by growth, but perhaps I have given enough to make a convincing case that multicellularity arose this way a number of times during the course of evolution. If we turn to multi-

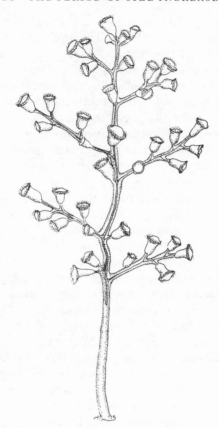

Fig. 9 A colonial ciliate (*Zoothamnion*). Note that all the cells are connected by a muscle thread which permits the whole colony to contract. (From Bonner, *On Development,* Harvard University Press, 1974.)

cellularity by aggregation, it is possible to see that here, too, there have been a number of independent inventions of multi-cellularity.

. . . . .

In some ways the most extraordinary case is found among lowly bacteria. There is a whole group of aggregating bacteria known as the myxobacteria (which means "slime bacteria"). They were discovered and described by Roland Thaxter at the turn of the century. He was the Ph.D. adviser to my adviser,

William Weston, so I have always thought of Thaxter as my intellectual grandfather. He apparently was a formidable man as well as a formidable scientist. Thaxter discovered two major groups of organisms that were not known to exist before. One was the myxobacteria and the other was a group of fungi that previously had been thought to be hairs on the bodies of insects and had been carefully described by some entomologists in the earlier literature as part of the insects. Not only did he discover that they were fungi, but he proceeded to describe some five thousand species, all of which he illustrated with his own beautiful stipple drawings. He apparently used to sit by the hour with a crow-quill pen and a large magnifying glass to put each stipple in its right place. I still imagine him in my inner mind with his stern, bearded face, his high, starched collar, giving each dot his personal attention. Nothing was out of place in his world, and perhaps for this reason he could see things others could not.

He found peculiar small fruiting bodies near his summer home in Kittery, Maine, and brought them into the laboratory to discover to his astonishment that they were not fungi, nor slime molds, but multicellular bacteria. He showed that they were rod-shaped bacteria which formed swarms, and clearly the cells multiplied during this stage; they were feeding and growing cells. After some time the swarms would form dense aggregates, which then rose into the air a millimeter or so high, and produced terminal, encapsulated cysts, each one of which contained many bacterial rods ready to germinate for another generation should new favorable conditions appear (fig. 10).

There is a suggestion that the stalk that causes the fruiting body to rise is produced by a portion of the cells that dies in the process of exuding the stalk, all for the good of the cells at the tip, which are encapsulated into the cysts. Should this prove to be the case, we will have another example of division of labor among cells in a prokaryotic organism. Not all myxobacteria produce elaborate stalks; a number of species form simple mounds in which each bacterial rod becomes transformed into a spherical, unicellular spore. In these simpler forms there is also evidence that not all the cells manage this

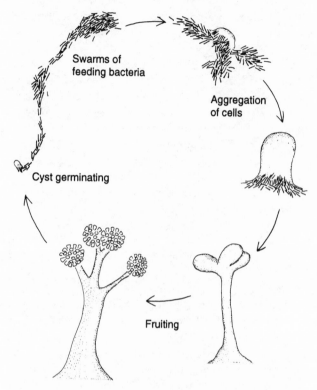

Swarms of
feeding bacteria

Aggregation
of cells

Cyst germinating

Fruiting

Fig. 10 The life cycle of the myxobacterium *Chondromyces crocatus*. A mature fruiting body (*lower left*) bears cysts, each of which liberates numerous bacterial rods, which accumulate in progressively larger groups, ultimately producing a new fruiting body. (From Bonner, *On Development,* Harvard University Press, 1947.)

happy fate. A number of them die, and again it is thought that their programmed death in some way contributes to the construction of the mound.

It is remarkable to think that slime molds have a life cycle very similar to that of myxobacteria, the main difference being that the cells of the former are eukaryotic amoebae instead of prokaryotic bacteria—again clear evidence that they must have invented their multicellularity independently. To go one step further, even within the cellular slime molds there are two

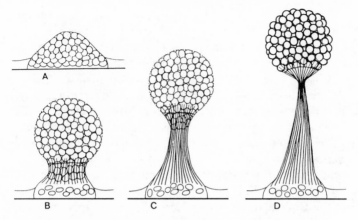

Fig. 11 The formation of a colonial ciliate. The cells aggregate and rise up on a stalk they secrete. Ultimately each ciliate cell becomes encysted. (From L. S. Olive, *Science* 202 [1978]:530.)

groups based on major differences in the structures of their amoebae. One group includes *Dictyostelium* and all the beautiful allies that are commonly used as experimental organisms today, while the other forms rather coarse and irregular fruiting bodies. The question is whether amoebae first became social and subsequently changed their cell structure, or did they follow the reverse sequence in their evolution. There is every reason to believe that the cell structure is of a more fundamental character and preceded their coming together in social aggregations. This again suggests that even within the cellular slime molds there have been at least two separate origins of multicellularity.

One final example is a remarkable recent discovery of a ciliate protozoan that aggregates to form a fruiting body. The individual cells of this organism look rather like stubby *Paramecia* and they live in the thin film of water in humus. After a period of eating and multiplying, they stream together to central collection points, exude some sort of stalk material so that they rise up into the air, and each ciliate cell then becomes encapsulated in a resistant cyst (fig. 11). In many ways it is remarkably similar to the fruiting bodies of the myxobacteria and slime molds; we can add ciliate protozoa to the list of organ-

isms that become social by aggregation. I should say parenthetically that these novel beasts were discovered by a botanist,
but so far no one has had the gall to call them plants. Perhaps
we biologists are becoming less chauvinistic!

Ciliates have always had a great appeal to me. They seem so
different from all other organisms, even all other protozoa, yet
in their fundamental processes they are the same: their differences are more apparent than real. On my first sabbatical leave
I decided to work on something as different from slime molds
as possible so that my view of life would not become too narrow. It was a happy decision for many reasons, but mainly because Professor E. Fauré-Fremiet invited me to his laboratory
at the Collège de France in Paris. We moved there with numerous small children and all their paraphernalia and by incredible luck found an apartment within walking distance of
the laboratory—and what a lovely walk, through the Luxem-
bourg gardens among other places. By far the most exciting
part for me was getting to know "Monsieur Fauré," as he was
known in the lab. He was a short man (I always thought he
looked a bit like Toulouse-Lautrec except that he was agile
and normally proportioned) and very vigorous at the age of
seventy. He was also a man of great charm and kindness.
Everyone in the laboratory adored him, which was very evident
the moment one entered it. He had a number of very able and
attractive young women laboratory assistants whom he called
"mes petites." Whenever I had a problem, one of his "petites"
was assigned to help me, and it was from them that I learned
that the French did science, at least at that time, in a way that
was quite new to me. When they made media to grow their
ciliates, they never measured out the ingredients—they took a
handful of this, and half a handful of that, and a pinch of this
and that, and then filled the flask with water. I had not realized
how American science had been totally based on the German
tradition of exactitude in every detail, where every ingredient
was weighed out to the last decimal point. In France the laboratory tradition came from cooking, in which a great chef relies
on the practiced, not to say inspired, eye; the laboratory was
simply an extension of the kitchen.

Monsieur Fauré was exceedingly kind to me—the young, eager American who would have been easy to neglect. He would take me collecting for ciliates in the country around Paris, and those forays were special treats for me. He knew every pond and ditch for miles around, and exactly where to have lunch. He would always apologize by saying that this inn was "tout à fait en campagne," but that country eating was a joy. He knew not only his biology in a profound way, but he was enviably learned in many other subjects as well. Perhaps that is understandable, considering he was the son of the composer Fauré, the grandson of the sculptor Fremiet, and the nephew of the poet and philosopher Sully-Prudhomme. He was exceedingly modest about his background and accomplishments, but he would not tolerate inaccuracies. On one of our excursions we stopped to look in at Port Royale, the stronghold of the Jansenists and the abode of Pascal. We followed a small group with a guide, and every time the guide said something, Fauré would say, "No, no, that's not quite right," and then gave a little explanation with the correct dates and details. The other tourists in the group looked stunned, and the poor guide looked absolutely black. I felt lucky we were able to get out alive. For all these reasons I have a special place in my heart for Monsieur Fauré and his beloved ciliates.

To return to the matter of becoming large by aggregation, there is another kind of coming together that is far more prevalent. It is not generally thought of as aggregation, but it clearly is exactly that. There are a number of forms that have a continuous mass of multinucleate cytoplasm which in various ways spreads over an area to feed and take in nutrients. This is the growth stage of organisms that live on dissolved substances and on small organisms (such as bacteria). They are scavengers of the microworld. When the conditions are right, all this nucleated cytoplasm is gathered together in clumps to form fruiting bodies. This gathering of multinucleate cytoplasm is functionally the same as the aggregation of separate cells.

One quite large group of organisms that lives in moist environments, such as humus or rotting logs on the forest floor, are the myxomycetes (sometimes called "true" slime molds in contrast to the "cellular" slime molds). The myxomycetes have a sexual cycle in which the spores emit a single cell, which is a sex cell, and two such cells of opposite mating types will fuse to form the equivalent of a fertilized egg. This cell will begin to sop up dissolved food, and the nucleus divides as new cytoplasm is made during the process of growth. Therefore, instead of a multicellular feeding stage, it will be a multinucleate one with all the nuclei moving around in a common cytoplasm. This so-called plasmodium looks like a viscous liquid and, like a giant amoeba, slowly crawls about seeking food. If the conditions are right, especially if there is sufficient nutriment and moisture, the plasmodium may become very large—the size of one's hand or even larger. It is not an uncommon sight to see a beautiful slimy glob of bright orange, glistening on the surface of a rotten log. When conditions become less favorable for feeding, the liquid mass of cytoplasm suddenly will form many small humps and, depending on the species, form numerous stalks which rise up into the air as delicate fruiting bodies (fig. 12). During this fruiting the nuclei undergo meiosis, getting ready for the fertilization (which will follow spore germination), and each nucleus becomes encapsulated in a resistant spore ready for dispersal. From the point of view of aggregation the process is basically the same as that found in cellular slime molds.

A functionally similar situation occurs in fungi, and now we are talking about organisms that are neither inconspicuous nor rare, but are prevalent all over the surface of the globe and consist of many thousands of species. Here the fungus spreads out through the ground or rotten wood—whatever will give the dissolved nutrients to the advancing and branching filaments of the fungus. All these spreading filaments will be connected, and again if the conditions are right they suddenly form fruiting bodies that will raise spores up into the air for dispersal.

Let us examine one very common and conspicuous example: the growth and fruiting of mushrooms in the soil. If a spore of an ordinary field mushroom lands in the soil and germinates, it will send out a filament that grows at the tip. It feeds by absorbing the minute amounts of sugars, amino acids,

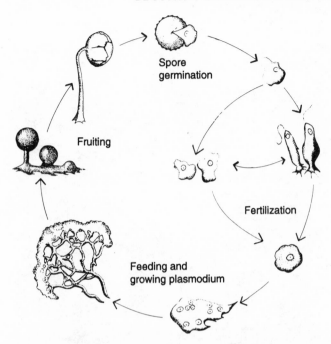

Spore
germination

Fruiting

Fertilization

Feeding and
growing plasmodium

Fig. 12 A generalized life cycle of a myxomycete. The spore
germinates, giving rise to a cell which, depending upon the
environmental conditions, is either an amoeba (under dry
conditions) or a flagellated swarm cell (under wet conditions).
After fertilization, the resulting cell grows into a large
multinucleate plasmodium that eventually turns into many
spore-bearing fruiting bodies. (From Bonner, *On Development*,
Harvard University Press, 1974, redrawn from Alexopoulos
and Koevenig.)

and other dissolved nutrients in the soil. As it does so, it syn-
thesizes more and more cytoplasm and its nuclei divide repeatedly, like the nuclei of a myxomycete. However, the multinucleate cytoplasm of a fungus is confined within the branching
filaments it has secreted. These filaments will not only spread
out evenly through the soil, but should they come near to one
another they may fuse; in this way the nuclei from different
spores, different genetic strains, may mix.

If the spreading of the growing filaments starts from one

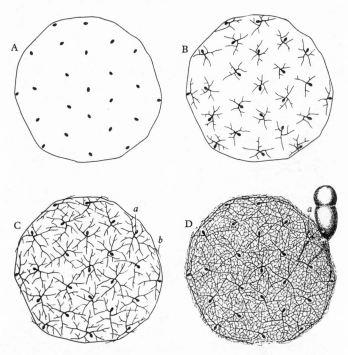

Fig. 13 The development of the small mushroom *Coprinus*. The spores germinate (A), and as the filaments take in nutrients from the agar medium, they grow and anastamose (B, C). Ultimately all the cytoplasm and the nuclei in the network of connected filaments flow into a small bud to form a mushroom. (From A.R.H. Buller, *Researches on Fungi*, Longmans, Green, 1909.)

point, it will spread out in a circle. Again, if the conditions are just right, at its outer edge it will show small buttons which are the primordia of the mushrooms themselves. It is at this point that aggregation occurs, for literally overnight there may be a sudden surge of cytoplasm from the feeding filaments into the primordial buttons. They will fill up like a balloon, for their cell walls are extensible and the result may be the sudden appearance of a large mushroom (fig. 13). In other words, mushrooms do not really grow overnight as is commonly thought.

They have done their growing for a long time beforehand, and their appearance and rapid expansion are the result of the aggregation of the nucleated cytoplasm that rapidly inflates them.

Because the feeding filaments form a circle as they grow outward in the soil, and because the mushrooms appear at the outer edge, there is often a perfect circular ring of mushrooms—it is a common sight. This is known as a "fairy ring," for it is thought that fairies sit on the circle of mushrooms and chat with one another. Undoubtedly this is so, but the scientific evidence is still wanting.

This same pattern is found not only in large mushrooms, which produce a prodigious number of spores to start the next generation, but in the simplest molds, where a single filament will rise a few millimeters in the air and a relatively small number of spores will be formed around the nuclei that are in the rising tip of the simple chitonous tube. Besides these two extremes in the size of the aggregation of nucleated cytoplasm, there are many intermediate forms giving a continuous gamut of sizes.

. . . . .

So much for my brief course in the natural history of inferior organisms. My hope is that I have convinced the reader that there is quite an extraordinary variety of independent, successful ways in which an organism can become large by becoming multicellular. We must now turn to the important question of why this is so, and what the advantages are from the point of view of natural selection.

I have already made the point that since there are clearly so many mechanically different and therefore independent inventions of multicellularity, there must be a strong selection pressure for it. At the same time, there is no loss of selection pressure for the smaller forms that continue to exist. The general argument is that multicellularity is an easy way for microorganisms to increase their size, and there is a continuous and significant selection pressure for a size increase. This brings the larger organism into another size world and it does not seem, at least in most cases, to outcompete the smaller, ancestral forms. A new, larger-size world is created by this invasion of a new ecological niche, enabling the large and the small to coexist.

While there is nothing false about these thoughts, somehow they are too general to be entirely satisfactory. One wants answers to the next level of questions—why is there a selection for size increase? Here the answer is not always the same for each foray into multicellularity. Ultimately we know it must depend upon reproductive success. As we shall see, this breaks down to two specific advantages. The first is to be more effective at obtaining food, which enables one to produce more offspring. The second is to increase one's ability to disperse spores or sex cells so that the progeny can land in new places rich in food and flourish there. As we shall see, these two related gains from size increase apply in different ways to all the examples of the invention of multicellularity that I have just described.

Not too long ago, when I was first considering the whole problem of multicellularity and preparing a lecture, I suddenly had an embarrassingly obvious flash of insight—one that probably has already occurred to the reader. All those organisms that have become multicellular by growth, by the cells that result from successive divisions sticking together to form a multicellular colony, are aquatic, while the organisms that get larger by aggregation are terrestrial.

If we first consider the aquatic forms that get larger by growth, we have a hard time inventing reasons why becoming larger might be an advantage. It is a favorite sport nowadays to make fun of just such hypotheses about why an animal or plant has adopted a certain shape or, as in this case, a particular size as well. They are derisively referred to as "just-so stories," with the implication that any hypothesis for the selective advantage of any particular adaptation might well be as wonderfully absurd as Kipling's explanation for the trunk of an elephant or the copious and loose skin of a rhinoceros. They would all fit the category of unsubstantiated fantasy. It is even possible that some things simply arose by chance mutation, and that they are selectively passive or neutral, so they have remained as they are because there is no selection against them. This could certainly be a possibility for size increase, and it should always be one of the alternative hypotheses one might make. Despite the invitation for a bit of derision from fellow biologists, inventing just-so stories seems to me neither a dangerous occupation nor a useless one. The only rule that must be obeyed is to make clear one is making a hypothesis; there can be no certainty until one has provided some real evidence. But where would we be in biology if we were not permitted to make hypotheses and allowed ourselves to be intimidated by the cry that we are committing the crime of just-so stories?

Unfortunately I have no good hypotheses as to why cyanobacteria have larger cells than bacteria, and why they make extensive filaments. It is conceivable that their large size has something to do with the efficiency of photosynthesis, especially in difficult and severe environments. But it is hard for me to guess what those advantages might be.

The same difficulty arises in considering the advantages of the large size of *Hydrodictyon*—or perhaps it is an even worse difficulty because they do not live in unfriendly environments. In the case of photosynthetic organisms, all one needs to do is be able to catch the sun, and all sorts of different shapes are equally effective in feeding off the sun's energy.

Perhaps the size increase in these organisms is a way to avoid predation by some organism that can easily eat bacteria or even small, single eukaryotic cells, but that cannot consume anything larger such as multicellular filaments or water nets. This is certainly a reasonable possibility. It could be that some but not all multicellular filaments arose for this reason; others may have appeared by chance with no selection for or against them.

Volvox also feeds by photosynthesis, but it has an added feature: it is motile. This means that it does not need to wait for the sun to come to it—it can, and does, swim towards the sunny spot in a pond. But there are many unicellular eukaryotic organisms that do the same thing. The only difference is that *Volvox* can swim faster because it is roughly true that the speed of an organism is proportional to its length, and *Volvox* is obviously very much larger than a unicellular form.

There is no evidence that becoming larger allows *Volvox* to be able to escape from predators more effectively than smaller, unicellular forms. That would only work if they could recognize a predator—a most unlikely possibility. Graham Bell has made the interesting suggestion that any organism above a certain size cannot be eaten by one of the many filter-feeders that live on small organisms in fresh water and salt. These would include clams, oysters, mussels, and many other invertebrates. The only difficulty with this idea for our argument here is that we are looking for the *first* forms to become multicellular, not the ones that have done so after the appearance of large, multicellular filter feeders. We can conclude only that the ancient causes of multicellularity in early earth history remain elusive today. It is even difficult to make convincing hypotheses.

Furthermore, there is another point that must be taken into account: there are attached or sessile forms among the ciliates and the diatoms, as we have seen. How do we explain these, especially if we assume that motility is such an advantage in finding light and other forms of food? Perhaps they live in places where the best strategy in finding food is to wait for the currents to bring it, rather than to rush after it. In the case of diatoms, which are photosynthetic, the advantage of being sessile versus free-living and motile is also puzzling and needs further study to discover what their relative advantages are in different habitats. Perhaps the forms fixed to the bottom can avoid being carried away to less favorable localities. The only thing we can be certain of is that both forms exist today and in their own way are successful survivors of evolution by natural selection. The reasons that this is so remain an intriguing problem.

Making hypotheses for the advantages of size increase and multicellularity in terrestrial organisms is easier to manage. To begin with eating, all the forms that have a multinucleate feeding stage, namely the myxomycetes and the fungi, can manage much larger items of food than single cells. This is because they secrete digestive enzymes into their immediate environment which dissolve large particles of food into their constituent small molecules, especially sugars and amino acids; these they can sop up through their membrane. Because they are multinucleate and large, they can secrete massive doses of these external enzymes and subdue large particles of food, far bigger than a single cell could manage. This has been called the "wolf pack" method of feeding, stressing the advantage in numbers and size to increase the range of foods available. This is in sharp contrast to the cellular slime molds whose small, uninucleate amoebae feed separately by engulfing bacteria. They are limited to bacteria and small yeast cells in the size of the food they can eat. For them there is no advantage in being multicellular at the feeding stage; it is helpful only when they have fruiting bodies.

In all these examples of terrestrial aggregation organisms, I have described how they invariably produce some sort of fruiting body which no doubt contributes to effective dispersal of spores that may colonize distant patches of abundant food. This general principle applies not only to microscopic fruiting bodies, but to those of larger fungi and even of higher plants of all sizes. In the latter case, seeds rather than spores are involved; one of the important functions of grasses, shrubs, and trees is seed dispersal. The mechanisms of dispersal often involve animals which swallow the seeds along with the whole fruit and spread them in their droppings as they wander about. Nevertheless, in some cases it is quite clear that the height of the tree is used to facilitate dispersal, especially when it is aided by wind. Who has not noticed that on many conifers the cones are produced near the top of the tree, thereby encouraging the greatest distance possible for the falling seed and the best chance of invading new areas where growth might be favored?

It is quite obvious that if an organism is confined to one spot it must have some way to send some of its propagating cells to other spots to colonize them. Its reproductive success depends on dispersal; without an effective system of dispersal it would go extinct, especially when in competition with other forms that are successful in getting their spores to all the fertile spots in the neighborhood.

Aquatic organisms must disperse for the same reasons, but they have much less of a problem. In the first place, many of the primitive multicellular aquatic forms, like *Volvox*, are motile and can disperse by swimming. Even stationary forms, such as the colonial ciliate *Zoothamnion* we saw earlier, give off a special cell which is freed from the parental colony and can swim off to start a new colony. Second, dispersal is effected in streams, lakes, and the ocean, by currents, which can carry any propagating cell great distances. For these passengers of the currents, motility is hardly necessary.

Terrestrial microcolonies, with their group of spores at the tip, do in some cases also rely on currents, but they are air currents. The spores of many fungi and myxomycetes are wind dispersed. Some myxomycetes even have a lacelike network of nonliving material between the spores that expands with drying, making the spores even more exposed and susceptible to the smallest waft of air. In other species, such as cellular slime molds, the spores are sticky and adhere to passing insects in the soil or humus, and in this way they are carried afar. If one looks to fungi, the methods of dispersal are cleverly devised, and wonderfully diverse. For instance, there is *Pilobolus* (not the dance group), which is related to the common bread mold and shoots its bolus of spores a considerable distance. It does this by holding the spores in an inverted cup and allowing the pressure to build up behind the cup so that it everts explosively when it can no longer resist the rising force. The spore mass is propelled some inches away; furthermore, the mass is sticky so that it will adhere to whatever it hits. I could go on with innumerable examples (and in fact C. T. Ingold has devoted an entire book to the subject), but I will end with bird-nest fungi as a final example. Instead of having a stalk with a terminal mass of spores, these fungi have tiny cups, with the spore masses at the bottom, looking for all the world like miniature bird eggs in a nest. At first this fruiting body seemed impossible—it was disobeying all the rules. However, it was discovered that these nests were splash cups. When drops of rain hit them just right, they would splash the spores upwards so that they could travel some distance. Besides providing an extraordinary example of a dispersal mechanism, this method also serves to illustrate that, like wind, rainwater in general can serve as an effective dispersal agent.

The fact that today we have so many dispersal mechanisms among primitive terrestrial organisms, and that those mechanisms differ so widely in their method of dispersal, is in itself very important. If we take a backward look at this bit of evolution we can come to two obvious conclusions. One is that there must have been an extraordinarily strong selection pressure among terrestrial forms for effective methods of dispersal; moreover, the response to this pressure has been extremely varied in that there has been a large number of independently devised dispersal mechanisms.

Success in reproduction also means finding food and devising effective ways of capturing and devouring it. This applies both to animals and to photosynthetic plants. Passive dispersal and active locomotion are the two ways of finding food, the former being characteristic of the terrestrial, aggregative forms. Note that none of the early terrestrial forms are photosynthetic. That process seems to be best accomplished in the open water where the sun's rays can easily penetrate. Aquatic forms can disperse actively or passively by currents, but their advantage in size only weakly contributes to more effective dispersal. Rather, their size increase by growth must either have other advantages, or there is no selection for or against it—it is a neutral character.

One particular point can be made about the evolutionary lines that led to higher plants and higher animals. Both clearly came from ancestors that became multicellular by growth and therefore had an aquatic origin. There must be severe size limits to what can be achieved by aggregation. Imagine an elephant developing by aggregation—an absurd idea from any point of view. But by growth, by the slow, progressive accretion of cells, it is possible to produce all the needed supporting tissues such as cartilage and bone, or wood in the case of large trees. A large mushroom seems to be about the upper size limit of aggregative organisms, and most are much smaller. Therefore, for each experiment in multicellularity there are two questions: (1) the survival of the ancestral form, and (2) its potential as a building machine that produces descendants, in the long term of evolution, who can achieve a very large size.

As the reader is by now no doubt aware, I have an inordinate fondness for grand ideas, even if they have sprung originally from the simple life cycle of a primitive slime mold. In my early days as an assistant professor at Princeton, a friend called me to say that he had told Professor Einstein about a film of slime molds I had made as a student, and would I be willing to show it to him. I was obviously thrilled and showed up with film and projector at the appointed hour. We had trouble finding a suitable screen but finally took a wall map of the United States and turned it around. After the viewing, Professor Einstein asked me if I would come into his study to discuss what we had seen. We were joined by the mutual friend who had arranged the meeting and Miss Dukas, Professor Einstein's secretary. We talked for some time, and what impressed me particularly was the depth of his questions. He wanted to know immediately the answers to all those questions that have been pursuing me all my life: How was the life cycle controlled so that it was the same each generation? Why does this kind of organism exist at all? Where does it fit in with other animals and plants? And he had many related questions. I wish I could have another conversation with him now; I have had much more time to think about these problems.

Professor Einstein was extraordinarily kind to me as a young beginner and seemed to have no trouble understanding my English. Occasionally he would stop to think, whereby those who thought he had not understood me explained in German what I had said. Each time he would flash furiously at them and ask them not to interrupt: of course he understood me! After we had finished and stood up to go, I told him that I knew the philosopher Alfred North Whitehead and had once asked him if he had ever met Einstein. Whitehead replied that indeed he had—under the most embarrassing circumstances.

Lord Haldane, a very forceful man, had invited both of them to dinner. After dinner he escorted them to his study and left them there alone, saying they must have so much to say to each other. He told me, "Both Professor Einstein and I are very shy men, and we had an excruciating time—neither of us could think of what to say." I asked Professor Einstein if his memory of the event was the same. He gave me a warm smile and said it certainly was—it was a painful evening indeed. "You see," he said, "I was never able to understand anything White-head had written, so what could I say?"

. . . . .

By considering early experiments in multicellularity, we are looking at early and varied ways of making the life cycles of single cells more elaborate. The part of each cycle in which the multicellular condition is attained is the part we call develop-ment. It is the building phase, the period of size increase, and we want to understand the mechanics of this development. Today this is one of the central problems in biology. In primi-tive colonial organisms we see not only the origin of multicel-lularity, but also the origin of development.

Chapter 4

BECOMING LARGER BY
DEVELOPING

MY PLAN in this book is to dissect the different phases of the life cycle and discuss them separately. In this chapter I shall concentrate on the period of size increase in any one cycle; in other words, the period of development. In the next chapter I will consider how that period of size increase can change and become progressively longer (or shorter) during the course of evolution; it will be a consideration of how the period of development can evolve over large periods of time and in many successive life cycles. Finally, in the remaining chapters, I will consider the period of adulthood along with all the remarkable things adults can do to be successful in reproduction.

Earlier I stressed the point that a life cycle generally has a single-cell stage, a requirement of sexuality, and that by increasing the size of the adults that bear the eggs and sperm, organisms are avoiding competition, or better, managing to co-exist with lesser forms; they rise above their competitors. (This is a rather grand oversimplification, for even the largest animals and plants can be felled by the smallest parasites or pathogens, but this occurs after the large organism has evolved.) The period of development from the single-cell stage will vary according to the size of an animal or a plant. For the most part, the first multicellular experiments discussed in the previous chapter on the origin of multicellularity are small, and the development is correspondingly simple and rapid. But if we look at the great range of organisms of all sizes, we see not only longer and more elaborate periods of development, but we see variety in its mechanisms. It is, after all, not surprising that a plant, with its stiff cell walls, should develop differently from an animal whose cells are bounded only by their membranes and are capable of moving about.

There are three basic, constructive processes that take place as an egg unfolds into an adult. One of them is growth—cell division (or, as we have seen, in some instances it is only nuclear division with a corresponding increase in a common cytoplasm). Growth means synthesis of new cell materials, for which energy is needed. Initially this may come from yolk in the animal egg, or from equivalent storage deposits of food in plant seeds. Once animal embryos achieve a certain size and have developed a digestive system, they are free to take in food energy on their own to complete their period of size increase. The same moment comes to higher plants when their first green leaves, capable of photosynthesis, appear. After that, all the food is used for the activities and maintenance of the adult, the very reason we need our regular meals each day.

Besides growth, there is another constructive process that is found only in animals: the movement of cells from one part of the embryo to another. These so-called morphogenetic movements play a crucial role in the shaping of animal embryos, just as they do in slime molds, where aggregation brings the cells together to form a slug, and then the slug undergoes cell movements which ultimately shape the final fruiting body.

The third constructive process is differentiation, which I have already discussed. It is where cells become specialized in dividing labor. This is a major event during development and one of the profoundly interesting problems of developmental biology. What decides that a certain cell will become a muscle cell or a liver cell, or a flower cell or a leaf cell, or a stalk cell or a spore? This is a question we will examine in some detail.

I am going to divide my discussion into two parts. First I want to consider the findings of "old" embryology; this was the information I was raised on. Then I want to go on to the "new" developmental biology. I will not simply chuck out the old and stick to the new because the new is based entirely on the old. The old developmental biology consisted almost entirely of descriptions and clever experiments. In this way the basic problems and questions were revealed. The new embryology is attempting to understand the molecular basis of those old questions using a splendid array of new techniques. It has meant a great leap forward, with all sorts of riches on the horizon, but it is a superstructure built on the older work carried out at the end of the last century and the early part of this one. Even today there are some who continue to do biological rather than molecular experiments on development. Since I am one of those, I am a firm believer that the experiments should continue in conjunction with the powerful new molecular methods. The old methods continue to serve the essential function of defining the problem, and they even play a key role in interpreting the greater meaning of the results from molecular experiments. In fact, recent studies show that there is not much of a gap anymore between the two, for it is only by using both that we can expect to continue to make significant advances.

I will not go back to the very early days of embryology, back to Aristotle and all the early debates, fascinating though they are. I remember being part of a Ph.D. examination in which the candidate was asked what were some of the earliest theories of the transmission of nerve impulses, whereupon he launched into a discussion of what Galen, the early Greek master of medicine, had to say on the subject. There was a stunned silence, for the question was certainly being answered correctly, but finally the examiner became restive and said, "I didn't mean quite that early—could you start in the nineteenth century?" Here I will follow the same course and begin in the late nineteenth century.

By that time there was an enormous accumulation of descriptions of the development of many different animal and plant species, and there was a full appreciation of the great variety of different methods of construction. This variety is not surprising if one compares the shape of a tree, a frog, an insect, and all the other differently shaped invertebrates and lower plants: how could there not be vast differences in the way they are built from a single cell? The excitement began with the use of experiments. On both sides of the Atlantic, two totally separate sets of facts seemed initially to be diametrically opposed to one another. In America, Dr. Conklin, who was beginning his Ph.D. thesis with Professor Brooks at Johns Hopkins University, became interested in the development of a mollusk found in the ocean at Woods Hole in Massachusetts. The cells during early cleavage were easy to follow, and he showed that each cell was destined to become a specific part of the larva. The whole plan of development was laid down in the beginning. Later he showed that if he removed specific cells, the larva simply lacked those cells or the tissues to which they belonged.

He used to tell us of his enormous excitement at this discovery, and his even greater excitement when he spent a Sunday with E. B. Wilson, who was working at the same time on the development of an annelid worm. Wilson later became America's—possibly the world's—most famous cell biologist. He was for many years a professor at Columbia University, and his lasting legacy is his famous book, *The Cell in Development and Heredity*, still read today with profit. When he and Conklin met they were thunderstruck to discover that both species, even though only distantly related, had exactly the same kind of development, which came to be known as "mosaic" development. The fates of all the cells were spelled out in a set of initial rules, like the commandments of Moses, and, unlike those commandments, they seem to be followed perfectly and in every detail.

Professor Conklin also used to tell another story which bears on this work. When he finally wrote up his discovery for his Ph.D. thesis—a landmark in the history of embryology—Professor Brooks, his mentor at Johns Hopkins, read it and was quite unenthusiastic. A few days later he returned it to Conklin, saying, "Well, Conklin, I've decided that this university has given doctor's degrees for counting words in ancient languages; I don't know what they might give you a degree for counting cells."

The other actor in this story was a colorful and imaginative embryologist in Germany named Hans Driesch. He found that he could cut up hydroids (which are related to sea anemones— freshwater hydra is a hydroid), and each portion would regenerate into a perfectly proportioned new individual. Even more startling was his work with sea urchin embryos. He found that if he cut them down the middle at an early stage, he would produce two miniature but perfectly proportioned embryos, complete in all their particulars. This so-called regulative development appears to be exactly the opposite from what Conklin and Wilson found in mosaic embryos, where two half-embryos would be the result. Human nature always creeps into these stories in very interesting ways. Driesch was so impressed by what he felt was a supernatural power in this embryo to restore the lost parts that he concluded there is no possibility of explaining regulation in any material or mechanical way. In fact, he became convinced that there must be some vital force, some spirit inside the embryo, that rearranged the amputated embryo—how else could one explain something so miraculous? When I worked in Conklin's laboratory one summer I asked him about Hans Driesch and his work, and it became quite apparent that their personalities had meshed no better than their discoveries.

It was not long before embryologists became reconciled to these apparently different kinds of development, and the question then turned to what the fundamental difference was between them—how could either one be explained? Driesch made a pronouncement, a dictum, that the fate of a cell is a function of its position in the whole. Therefore if a cell finds itself on the outside of an embryo it automatically becomes an epidermal cell. This was how he perceived each half-embryo was able to reconstitute itself into a miniature whole embryo: the cells on the cut side suddenly became outside cells and as a result differentiated into epidermal cells. This explanation, along with his dictum, strongly influenced the course of early twentieth-century embryology.

Driesch's impact is well illustrated in the fascinating story about H. V. Wilson and reconstitution in sponges and the interpretation of this process. In the first decade of this century, Wilson was working at a marine biological station on the coast of North Carolina, where, for some reason, he decided to push a small sponge through a very fine silk cloth which acted as a sieve. It had the effect of separating all of the sponge's cells, causing a rubble of cells to settle on the bottom of a dish. To his amazement and delight, the rubble slowly began to reorganize itself after a few days, and after seven days he had small reconstituted sponges. Driesch's influence was so strong at the time that Wilson interpreted his experiment to mean that the trauma of being pushed through the bolting cloth had caused all the cells to revert to their embryonic condition, and that when the separated cells came together, those on the outside became the covering cells and those on the inside became the flagellated cells that produce the internal water currents which bring in the food.

That this interpretation was not correct was shown conclusively by Julian Huxley in the early 1920s. Huxley was the grandson of T. H. Huxley and the brother of Aldous Huxley and is well known for his work in evolutionary biology. He worked on a species of sponge in which it is possible to get pure isolates of the flagellated cells, free of any of the other cell types. If Wilson's interpretation had been correct, this collection of one cell type would have produced a perfect, small sponge, but it did not—it simply remained a group of flagellated cells. There was no reversal of differentiation. As a result, Huxley suggested that in this case the fate of a cell was not a function of its position, but the reverse: its position is the result of its fate. In other words, in Wilson's experiments the cells did not change their differentiation and the flagellated cells migrated inside while the covering cell migrated to the outside in the initial clump of disturbed cells. This explanation holds today, and this kind of cell sorting is known to occur not only in sponges but essentially in all animals with motile embryonic cells.

The explanation has been provided in a wonderful insight by my old friend and colleague, Malcolm Steinberg, who showed that each cell type will stick to its own type or other types with different degrees of vigor, and these differences in the forces of adhesion determine the pattern of the sorting out of the cells. If flagellated cells adhere to one another more strongly than they adhere to covering cells, and more strongly than the force existing between covering cells, one can exactly predict the outcome: the covering cells will move to the out-side and the flagellated cells move to the inside. Steinberg has done most of his experiments on vertebrate embryonic cells, showing that the theoretical predictions of the results of these purely physical forces can be clearly substantiated by careful experiment. Today numerous laboratories are isolating and studying the cell-surface molecules responsible for these differ-ent forces of adhesion, putting the whole subject on a firm mo-lecular basis.

The reader may have assumed from what I have written so far that Conklin was right and Driesch was wrong. Happily that is not the case. Both of them were right, and mosaic development, sorting out, and regulative development *all* occur—often in one organism. Take, for example, Conklin's ascidians, which are so beautifully mosaic. Yet many ascidians that develop this way have, later in life, extraordinary powers of regeneration. In the colonial ascidians, which are attached to one another by a rootlike tube of cells, any portion of that tube can produce an entirely new adult if a piece of it is cut off. In one life cycle, the organism starts off strictly mosaic and ends up strictly regulative. Among all multicellular organisms there are an amazing number of permutations.

I became deeply involved in the problem because slime molds also reflect a mixture of mosaic and regulative develop-ment. Originally, as with sponges, we assumed that slime molds were totally regulative. Kenneth Raper had shown in his early work that if a migrating slug is cut into fractions, each segment would, under the proper circumstances, produce a normally proportioned fruiting body. If the cut portion comes from near the anterior end, all the cells are presumptive stalk cells, yet clearly those cells at the posterior end of the segment become transformed into presumptive spore cells—there is a forced intraconversion between the two cell types depending on where the segment comes from.

In the late 1950s, when I was a visitor in C. H. Wadding-ton's genetics department at the University of Edinburgh, I made what seemed to me a surprising discovery. (Waddington was a distinguished embryologist who did much to bring devel-opment, genetics, and evolution together.) I used mutant cells as a marker and found that the mutant cells were not uniformly distributed in the slug; I had expected this since they were randomly distributed before aggregation. I followed up with experiments using cells marked with vital dyes, and it was clear that at the end of aggregation there was a great mixing of cells. This observation was not only confirmed by many others, but it was greatly extended, so that we now know that even before aggregation, cells lean in either the spore or stalk-cell direction. Furthermore, we know some of the causes that determine these tendencies. The cells that are beginning to make stalk proteins sort out so that they lie in the front of the slug, and those that are forming spore proteins sort out to the rear. But these initial steps in the spore or stalk direction are reversible, as Raper's old cutting experiments showed. So here too there is a mixture of "the fate of a cell is a function of its position," and the reverse, all in one organism. Let me add that working in a Scottish laboratory had some distinct advantages for experiments with slime molds. The room temperature was around 68° F, which is ideal; but the central heating was turned off at noon on Saturday and put on again first thing Monday morning. All my slime molds therefore stopped in their tracks Saturday afternoon, and did not resume activity until midmorning on Monday. Everyone had the weekend off.

There is much more to the old embryology than I have included here, but I must resist the temptation to go on and on about one of my favorite subjects. Let me just mention that an important process is that of induction, during which one part of the embryo sends a signal to another to develop in a particular way. This process was first put on a firm footing by Hans Spemann in Germany in the 1920s, and chemical signaling continues to be a central subject in modern embryology, as we shall see.

．　．　．　．　．

The big step in the "new" developmental biology is in the linking of development with genetics. It was always understood that genes "acted," and when they did they were responsible for the determination of "characters." This is the basis of the classical work of Gregor Mendel in the last century, a subject that was brought to its first peak in the work of Thomas Hunt Morgan and his collaborators at Columbia University in the 1910s and 1920s. They showed that the genes on the chromosomes were in a linear order, and for the first time it was made clear how they could be shuffled and rearranged during meiosis so that the egg and sperm would have new gene combinations. It was also the first appreciation that one gene could mutate and that this altered gene would produce a new trait. For this work Morgan received the Nobel Prize, and of his own accord he shared the money with the remarkable scientists he had gathered in his laboratory. Many years later when I was a freshman at college and Morgan was the grand old man of genetics, his nephew, a friend of my parents, gave me a letter of introduction. I was spending the summer at Woods Hole at the Marine Biological Laboratory, where I was taking a summer course, and where Morgan was as well. I sent the letter on to him, but received no reply. I checked back home, and my mother assured me that it was socially correct to call on him anyhow. This I did with a combination of terror and teenage crust, and was rewarded with a wonderful evening with the Morgan family. They were playing bridge when I arrived, but my presence did not stop the game. They finished the rubber because, as Morgan explained to me, he was winning. Meeting one's childhood heroes often turns out to be a great disenchantment, but this was not so for me. I had more spring in my step for a long time afterwards.

The rise of genetics in this century and the birth of molecular genetics is, as everyone today knows, one of the most impressive success stories in science. Because I was young enough to have met Morgan, I have been able to witness the entire drama unfold around me all the years of my career. I was never part of it because there is nothing about me that allows me to think like a geneticist. It requires a special kind of brain, as it does to be a mathematician or a musician, and I am quite hapless in all these mental activities. However, the significance of the amazing advances in genetics did slowly seep into me, and it is now obvious that the future of developmental biology lies in the hands of modern molecular genetics. I will very briefly outline what those advances have been and then will show their significance for developmental biology, both for the present and the future.

Everyone has heard of DNA and knows a few things about its magic. For a variety of reasons it is often mentioned in the newspapers. There are now ways of changing the gene structure of some organisms so that, for instance, there is the possibility of developing disease-resistant crops. It is also possible to produce some important biological substances that are impossible to obtain in quantity in any other way, such as a blood-clotting factor for those suffering from hemophilia, or a growth hormone to cure a particular kind of dwarfism. And recently there has been much interest in using DNA, which is unique in every individual, for "fingerprinting" to find criminals. Vast sums of money are being sought for research so that the entire DNA structure of human beings can be known. This could lead to a greater understanding of genetically determined diseases and perhaps to a way of preventing them. Here I want to say some very simple and straightforward things about DNA and leave out the great array of details that comprise our present knowledge. I will stick to the important principles.

DNA is made up of four kinds of molecules (small molecules called nucleotides, or simply "bases"), and these are strung into a very long chain. In a small organism such as a bacterium, the DNA in each cell is in a circle containing about ten million nucleotides; human beings have as many as a billion, an enormous number. DNA generally does not exist as a single strand, but as two strands in which the nucleotides of one complement those of the other. The bases then become "base pairs," and certain bases can pair, or chemically combine. This double strand is the famous double helix of James Watson and Francis Crick, which they first described in 1953; for this work they were awarded the Nobel Prize. Part of the reason this discovery is so well known is because Watson published the amazing account of his research in his best seller, *The Double Helix*. This was the first time in the history of science that anyone had both the skill and the courage to write about an important discovery with the frankness and bald honesty of a first-rate novelist. Perhaps most remarkable of all is that his portrait of himself, the central character of the book, is hardly flattering, which gives him wonderfully free license to say prickly things about almost everyone else in the plot too. I am told by a friend who saw the first draft that the published version is mildness itself by comparison. Even though Watson has led the way in scientific biography, he has not begun a trend—few scientists have the internal toughness to carry it off.

The key feature of the double helix is that one strand of DNA can, by separating from its partner, make a new complementary strand. Free bases surround the open strand and attach to their complementary base on the single strand. This process is a replication, that is, new DNA can be made from old DNA in an exact, template, complementary copy. Here lies the key to genetic inheritance, for genes are sections of the long strands of DNA, and these can be copied with great precision and handed down from one dividing cell to another, or from one generation to the next, for the DNA from each parent will be in the egg or the sperm.

One of the main elements of Darwin's theory of evolution by natural selection is that organisms vary, and this variation must be inherited. The question we now ask is: What does this variation mean in terms of DNA, since that is the chemical which makes up the genes? The simplest way for a change, a variation, to occur is through the accidental change of one nucleotide. This kind of error, or mutation, is a common occurrence.

Another way of producing variation is to reshuffle the DNA during meiosis, that is, during the process of forming the egg and sperm. Part of this reshuffling is the result of the ability of DNA to break apart and stick together at the broken ends with considerable ease, and much of the most recent work on variation hinges on this particular capability. For instance, it is possible that sections of DNA will drop out completely (and the remaining severed ends unite), and such deletions will often produce a visible mutation, although the chances that some essential DNA is lost and the organism dies during early development are considerable. It is also possible that small bits of DNA, called plasmids, will be free in the nucleus of the cell and will become added to or spliced into the chromosomal DNA so that the organism has a new gene that may affect its structure and its appearance. Furthermore, there is evidence that in some instances these small circular fragments of DNA can pass from the cells of one organism to another, bypassing the normal meiosis and being transferred as separate entities during fertilization.

Even more remarkable are the genes that move around from one place in the DNA strands to another. The relocation can also produce a mutation and change the characteristics of the adult. This phenomenon was first demonstrated some time ago in some elegant genetic experiments on corn by Barbara McClintock, but it was not generally appreciated or understood until we began to see how DNA strands could detach and reattach with such ease. This is a remarkable story, for McClintock was, for a long time, quite unappreciated except by a few corn geneticists who admired her work but could only puzzle at the strange results and conclusions she was producing. One reason for her relative obscurity, except among the cognoscente, was that her papers are extraordinarily difficult to understand; another no doubt was that she was a woman, in a period when they were far less listened to in science than they are now. But then a wonderful event occurred. The molecular geneticists discovered mobile units of DNA that could wander from one part of the genome to another, and suddenly all of McClintock's work was fully appreciated for what it was worth: an extraordinarily clever genetic demonstration of something that could now easily be demonstrated by molecular means. Instantly she became a shining prophet, and justifiably so, having anticipated an important field of genetics years before it come into full flower. The final happy result was that she received the Nobel Prize in 1983 for work she had done from the 1930s to the 1950s.

It is important to remember that despite all this evidence for the mobility and rearrangement of genes, these are not just a mass of chaotic processes but are often carefully controlled.

Like all mutations, some of them are chance events. Chaos is avoided because mutations occur at such a low frequency that those producing disadvantageous results are culled out, for the organisms that contain them will not survive. In other cases there are protein enzymes or even segments of DNA that regulate the degree of mobility and exchange of the genes. This control fits under the general rule that, for natural selection to effect change, there must not be too much or too little variation—there must be a compromise.

It is time now to ask: What do genes do? How do they give off instructions to produce a new organism each generation? Chemically, DNA produces RNA which in turn produces proteins. Francis Crick called this the central dogma, with the idea that the three steps could go only in the direction indicated above. We now know the dogma is not quite true for certain RNA viruses (retroviruses)—some of which cause serious illnesses, including some cancers and AIDS in human beings— that will, as simple RNA strands, give rise to DNA by template replication. The newly created DNA engineers the future success of the virus at the great expense of the host.

DNA produces RNA right on the chromosome in the nucleus. It is a template replication called "transcription." The process is possible because DNA and RNA are chemically quite similar; they differ chiefly in the kind of sugar they have attached to the four bases. The resulting "messenger" RNA separates from the parent DNA and joins a curious, contorted molecule called the ribosome. This complex lies in a sea of amino acids and in a number of small RNA molecules called "transfer" RNA. There is a transfer RNA for each of the twenty kinds of amino acids, and it combines specifically with its own amino acid. This double molecule joins the ribosome-messengerRNA complex, and miraculously the transported amino acids are then zippered into a protein chain which corresponds to the sequence of the code in the messengerRNA (and which originally came from the DNA code of the gene).

All of which brings us to the whole subject of the genetic code. How does the DNA determine the sequence of amino acids in a protein? Proteins have twenty amino acids to choose from, and they can be arranged in any sequence in chains that vary greatly in length. The result is an enormous variety of structures for different proteins, a key property that makes them so essential to living organisms. All living processes are governed by a vast array of different kinds of proteins, and all are determined by the DNA. But how can this be, for DNA has only four bases, falling far short of proteins with their twenty amino acids? The question literally turned out to be a problem of cracking a code, and it was shown that an amino acid was determined by a sequence of three bases. For instance, in order to produce the amino acid alanine, three bases are required in the messengerRNA in a specific sequence—guanine, cytosine, uracyl. When these are brought to the surface as the messengerRNA is held by the ribosomalRNA, an alanine molecule will attach onto the growing chain of amino acids. In this way the four bases of DNA (and RNA) can be "translated" into one of the twenty amino acids. Note that in the translation the vocabulary has been amplified—from four bases to twenty amino acids, which explains how the enormous variety and richness in the diversity of the proteins in our bodies are possible.

· · · · ·

If we now aim all this powerful information and the tools that go with it at development, we will quickly see that we are under the foundations of the old developmental biology. We know that in certain cells a particular gene, at a particular moment, will synthesize its protein, and the protein will play a part, make one step, in the development of the organism. This can be vividly demonstrated by labeling the protein and seeing where it lies in the cell and in the embryo. Furthermore, it is possible to find not only the sequence of nucleotides for a particular gene, but also the sequence of amino acids in the proteins (since we know the code). By a combination of methods one can find out not only what cells produce which proteins, but how the gene products of one cell will affect the activity of genes in neighboring cells. Sometimes the products stimulate neighboring cells, and in other instances they inhibit them. The ways local gene actions within the embryo can affect other regions is rich and varied. For some steps in the development of some embryos, we now know how genes bring induction about—how one cell or group of cells can signal another to follow a particular course.

The two organisms which have received most attention in these kinds of studies are the fruit fly, *Drosophila*, and the soil nematode, *Caenorhabditis*. Their genetics are now well known and understood. It is easy to make genetic crosses in them using a large array of mutants, and through the analysis of the mutants it is possible to isolate the genes involved, find their function in development, find where they are on the chromosome, and then, as I have just said, identify their nucleotide structure. In the case of the nematode worm, which appears to have a set, rigid, mosaic development, recent work has shown that in the final stages of differentiation the fate of a cell is by no means fixed and absolutely predestined, but is very much decided by communication with its immediate neighbors. By using fine laser beams, one can easily kill a particular cell, and if that cell is gone, neighboring cells often alter in their final differentiation so that one of them can take its place. Therefore, even a mosaic development retains some degree of regulation.

The fruit fly also shows this kind of communication among developing cells that are touching one another, such as in the formation of the fine structure of their eyes. Particular attention is presently being paid to the early development of the embryo. A number of genes are now known to establish the characteristics of the front and hind end, and the back and the belly, and ultimately the segmented body plan characteristic of all insects. There are two especially interesting features about these early genes. One is that many of them reside and are expressed in the egg as maternal messengerRNAs. In other words, the early embryo is told by its mother what first steps to take before it can make its own proteins.

The second interesting feature is related to the first. These messages are not placed everywhere in the embryo, but at one end. This is because in the mother's ovary the nurse cells which give the RNA messages for the future offspring are at one end of the growing egg, and the mother's instructions are imparted in this asymmetrical fashion. The early embryo has no cell walls but is multinucleate like the plasmodium of a myxomycete. As a result, the proteins which are made at one end can freely diffuse across the inside of the early embryo, producing concentration gradients of that protein. This means that the nuclei, and ultimately the cells of the embryo, will be surrounded by different concentrations of this signaling protein. To make matters more complex, there can be considerable interaction between these gene products, and ultimately they map out a perfect larva.

· · · · ·

I am quite aware it is time to pause for the reader to catch her or his breath. Like Mark Twain, I have been leaking too many facts. The difficulty is that if I want to give any understanding or feeling for what modern developmental biology is today, and its enormous importance for the future of biology in general, a certain level of information cannot be avoided. I have a clear recollection of a lecture I gave to freshmen and sophomores in my first year of teaching. I tried hard to explain what seemed to me an important idea concerning the relation of the parts of a plant to one another, and in particular Goethe's ideas on the close relationship between leaves and flowers, in what he called the metamorphosis of plants. In what undoubtedly must have been an inspired lecture, I showed how crucial the influence of *Naturphilosophie* was on the evolution of biological thought at the end of the eighteenth century. After the lecture a student approached me, filling me with hope that I had reached at least one person. He asked me how long it would take to die if one cut someone's jugular vein. I resisted the temptation to experiment on him on the spot, and solved my problem by never giving that lecture again.

· · · · ·

Many people, myself included, have been quick to point out that the process of genes producing proteins is obviously a crucial, central component of development, but it is only part of the story. That alone could not account for an adult animal or plant or slime mold. Even if we confined our attention solely to genes and their products, we would still need to see how they can produce a consistent pattern and consistent proportions from one generation to the next. In a multicellular organism each cell has the same genes. They cannot all be turned on in the same way at the same time in all of the cells—this would produce neither cell differentiation nor pattern. Somehow a way must be devised to make regional differences so that the genes expressed in one part of the embryo are different from those in another, and this difference must be rigidly controlled. In the case of the front/hind end axis of the fruit-fly egg, the polarity of the embryo was imparted by the mother leaving specific messenger RNAs at one end of the maturing egg in the ovary. However, that is only a small bit of starting information, and not all eggs are born with a polarity imparted by their mother.

Besides the genetic blueprint or program for development, there are other factors which play a very significant role. These factors are primarily physical forces that impose or produce certain conditions which can guide morphogenetic movements and produce pattern. There are those who would go so far as to deny any role for the natural selection of genes; these so-called structuralists insist that there are certain properties of matter which guide both development and evolution in some way that seems mysterious and elusive to me. For some the desire to bash Darwinism is strong, possibly because it is the accepted dogma, but what is substituted is something vague and unsatisfactory. It has not been possible to find and identify those structural forces that are supposed to direct development and evolution.

It is obvious from what I have said so far and have repeatedly emphasized in earlier writings, as indeed have many others interested in developmental biology, that development does not consist only of the synthesis of new proteins, but that many other subsequent processes play a vital role. Let me briefly outline some of these.

First, the proteins synthesized through gene action are often enzymes which control or catalyze specific chemical reactions. These reactions will produce another set of chemicals—either small or large molecules—that can play a further role in guiding development. The second string of chemicals can be signal molecules of various sorts, such as hormones or embryonic inductors. They may stimulate some further chemical responses or, in some cases, inhibit some other secondary or tertiary reaction that is taking place. During development there are complex networks of chemical reactions that were initiated by the genes and their initial products but will later be largely divorced from the genes which gave birth to them and will carry on an interacting, interlocking life of their own. The chemical reactions and pathways, sometimes extremely complex, will be the product of the properties of the substances themselves. The structure of these molecules permit the great sequence of chemical reactions to occur, but the original instructions for this great chain of chemical events come from the genes.

These sequences of chemical reactions do not, in themselves, explain the pattern of a developing organism; something more is needed. One answer to the problem came from mathematicians who showed that by the combination of chemical reactions and physical forces, such as diffusion of the molecules and the flowing properties of liquids, theoretically all sorts of patterns can be produced. It is not completely clear to me how these mathematical models got their start. My first introduction to them came from the books of N. Rashevsky in the 1930s and 1940s. He is never cited in the literature today, possibly because his formal mathematical formulations seemed far removed from empirical observations, and no doubt he was rather extreme and immodest in his claims for "mathematical biology." I never met him, but I understand he was a colorful personality (with a flowing red beard) who used to start his lectures by saying that physics was in a terrible mess until Isaac Newton straightened matters out, and that biology was in a similar chaotic state until, fortunately, he came along. Perhaps this was enough to make biologists look in another direction, or perhaps he was before his time and the world was not ready for his ideas.

The person able to put the reaction-diffusion approach in a form that was palatable and acceptable was Alan Turing, in his well-known and influential paper on biological pattern published in 1952. Turing was a brilliant mathematician who, as a young man, cracked the German code in World War II. After that he became one of the founding fathers of computers and their artificial intelligence. Because of his 1952 paper he is considered one of the key figures in biomathematics. Yet his life was filled with tragedy and became the subject of a biography and a fine play about him. He was an unabashed homosexual during the twilight period when homosexuality was still illegal in Britain. As a result he was arrested and forced to receive masculinizing hormone treatments, which slowly undermined his health until he finally committed suicide at what was still an early age.

In his 1952 paper he suggested that if one had diffusion gradients of key substances that were important to development (he called them morphogens) and if those chemicals reacted in various ways, not only in their production and destruction (source and sink), but also in their ability to activate or inhibit some developmental process, then from this reasonable set of conditions one could generate all sorts of patterns which are indeed found in organisms. For instance, in his paper he showed how one could get standing waves of a morphogen that could account for the position of hydroid tentacles around the mouth. Turing's initiative has now been greatly expanded, leading to two trends: one which pursues his "reaction-diffusion" idea, and another which takes into account the physical properties and the flow dynamics of the cells. Apparently the mathematics are quite similar, and both approaches have permitted us to develop new ideas of development that, although hypothetical, have been stimulating and rewarding. It has led, among other things, to a great search for morphogens in developing systems. Part of the problem so far is that more morphogens have been discovered than are needed for the simplest mathematical models; but this is to be expected and it means that the models and the experiments must be refined together. If we now try to see these reaction-diffusion explanations of development in terms of gene instructions, we are forced to the conclusion that the steps between the immediate gene products and the pattern are numerous and the path is tortuous.

Another way genes can produce secondary effects of enormous consequences is by producing molecules that lie on the outside of the cell membrane and affect the adhesiveness of the cells. This is perhaps most obvious in higher plants where each cell is surrounded by a rigid cellulose wall, and the cellulose box of one cell will adhere tightly to those of its neighbors as the cells mature. It is this property that allows trees to grow to such enormous heights, despite the combined effects of wind and gravity. It is not the genes that hold the cells together in this rigid structure, but their products, such as cellulose and many other polysaccharides and proteins, which give the rigidity of the walls and the glue that cements them together. There is a series of gene-initiated chemical reactions which produce physical properties of the cells that are for reading. For example, the older parts of a tree are so rigid that they cannot grow, so growth in plants occurs in softer growth zones such as the apex of the root and the shoot or the cylindrical cambium just inside the bark which is responsible for the increase in the girth of the tree.

We have already seen that in animals the adhesion between the cells of an embryo plays a significant role in the distribution of cells during embryonic development. I gave as an example the reconstitution of a sponge after all its cells had been separated into a rubble. The external covering cells migrated to the outside and the flagellated cells to the inside, and they did so because their motile cells had different surface adhesive properties. Their ultimate rearrangement was not due to a blueprint in the genes, but because the genes had endowed the surfaces with different adhesive properties. As Steinberg showed so elegantly, these differences in the adhesive properties of the various cell types were in themselves sufficient to account for the ultimate pattern. Again the final result is some distance away from the original gene products, and the final pattern can be achieved only by means of the different adhesion properties of the motile cells and by the physical consequence of these differences, which was originally specified by the genes.

One of the most surprising recent findings is that electrical fields may play a part in development. The idea of bioelectricity is an old one that had not been well received because it was tinged with crackpottery—for example, with the claim that electricity is the basis if not the secret of life, and all sorts of other nonsense. It was well established that there were voltage differences between different parts of an organism, but that these differences are in most cases not themselves a cause but the result of chemical activity occurring on different sides of impermeable membranes.

It was, therefore, a great shock when Lionel Jaffe in the 1960s showed unequivocally that developing systems not only had voltage differences between parts, but generated electric current of sufficient magnitude to make a positive, causative contribution towards development. He showed this with the germinating egg of rockweed, a brown alga that lines the tidal zones of our coasts. Rockweed has a beautiful spherical egg whose first sign of development is the bulging of an incipient root—or more properly, rhizoid—from one side. It is possible to control with light on which side of the egg the rhizoid protuberance will appear: it will always grow out from the dark side, the side away from the light source. Jaffe put a number of eggs in a small glass capillary tube and illuminated it from one end so that all the rhizoids grew out in the same direction. He put an electrode at each end of the tube and, behold, he was able to record a steady increase in the current passing down the whole tube as the rhizoids developed. By dividing the total current generated by the number of developing eggs in the tube, he was able to estimate the amount of current generated by each egg, which turned out to be considerable. There was enough to account for a phenomenon that had been observed in the rhizoid tip: a zonation of different components, especially proteins, that occurs as the tip develops and grows. The amount of current in one egg is sufficient to separate out proteins and other large, charged molecules (for each protein responds differently to the electric charge, depending on how many charge groups it possesses from its exposed amino acids). In other words, the current is sufficient to produce a pattern in the rhizoid.

By means of some ingenious experiments, Jaffe then went on to show how the current is generated. He showed that the flow of charged calcium ions enters the rhizoid tip and passes through and out to the other side of the egg. This is achieved by proteins at the cell surface that actively transport the ions across the cell membrane—from outside to inside at the rhizoid end, and the reverse at the opposite end.

In this case also the end developmental effects seem far removed from the immediate gene actions. Again the key proteins, as for instance the ones involving the active transport of the calcium, are direct gene products, but the current they generate is a step beyond, and the ordering of the rhizoid proteins into zones is still another step removed. The genes do not do all the work but create the conditions in which a series of events will inevitably occur.

Stuart Newman has stressed a very important point about the relation of genes to the various physical forces that play a role in development: if physical forces take over and consistently cause specific developmental steps from one generation to the next, certain of the steps may acquire genes which also govern those steps. They are not necessary, but seem to seep in unavoidably, a bit like my old chemistry teacher at school who used to wear both a belt and a pair of suspenders. This does not mean that genes which govern pattern are necessarily selected for. They could be mostly neutral mutations for which there is no selection—for or against; they are neutral mutations that have seeped in simply because there is no selection against them. Newman suggests that they also could have some advantages by stabilizing the developmental processes and thereby protecting them from the vagaries of environmental fluctuations, which could easily affect some of the purely physical forces.

This process of genes infiltrating or becoming assimilated to reinforce something that is already occurring is called the Baldwin effect. J. Mark Baldwin, who produced these ideas when he was at Princeton University at the end of the last century, was ignorant of genetics (as was the whole world at that time), but he was interested in the idea that evolution in animals could be led by behavior, and that a repeated, consistent behavior could become inherited. This is basically the same idea which is being suggested here for developmental processes. Later we shall return to this point, because I think it is an important and very general principle of biological evolution.

One final element in development deserves a brief discussion here: genes do not produce proteins at all times, nor do they produce them only selectively in certain environments. This latter external factor plays a key role in development, but there is yet another controlling factor: the matter of timing. There seems to be some sort of internal clock which decides when particular genes will be expressed.

Perhaps the first to state this clearly was J.B.S. Haldane in 1932 in a well-known paper. Haldane was an imaginative and brilliant British biologist who made the explicit point that there are a number of genes whose effects appear at different times during development and later during the life cycle. As a young man I gave some lectures at University College in London where he was professor of genetics, and after the first lecture I met him in the "gents." I hardly knew him then, but as I was washing my hands, he suddenly boomed at me: "Bonner, we don't make jokes in lectures in this country." I nearly slid into the sink, but had enough strength to say, "Those weren't jokes—I was just nervous." This was the beginning of an interesting (and never dull) friendship with a remarkable man, for he had wonderfully original ideas about many aspects of biology.

Haldane pointed out that there are mutations which appear early in development and cause defects in cleavage or gastrulation. These mutants are invariably lethal and development stops abruptly at an early stage. Many mutations appear at later stages of development and are viable. In fruit flies or human beings they may involve such characteristics as eye color, hair or bristle color, or in the flies aberrations of the structure of the wings; there are innumerable other examples. In human beings there are mutations which only appear in old age or during maturity. I already gave the example of the miserable genetic disease, Huntington's chorea, that causes a massive and lethal degeneration of the nervous system and strikes only after the age of forty.

As we have seen, some animals and plants, such as mayflies and Pacific salmon, go so far as to be programmed to die after reproduction. There is a tree in the tropical forests of Panama that grows over many years to a great size without reproducing; and then in one season it will produce a huge mass of seeds, after which it immediately dies, providing open space in which some of its seedlings can grow. In these cases there seems to be a genetically determined death.

Finally, Haldane showed that some genes have their effect on the next generation. At first this might seem impossible, but there are a number of mutations which occur during the formation of the egg in a female, and therefore have their effect in the next generation. There is, for instance, a gene called "grandchildless" in the fruit fly; it gives rise to a change in the developing eggs, producing individuals that are sterile. Such a gene would be selected out immediately, but there are others that are viable and exist in nature. One of the best-known cases is that of coiling in pond snails. They may either coil to the right or to the left, and this character is genetically determined. However, it exerts its effect during the formation of the egg, so that an individual carrying the genes to produce a right-hand spiral may itself be left-handed, for the genes can have their effect only on the animal's eggs, not on itself. Because the genes act on the developing egg, their effects are not seen until the next generation.

. . . . .

If we take a panoramic view of all we know about development, we can see that its mechanics can be explained only by a great combination of factors. There are the genes which make the initial messengerRNAs and proteins; they start the process off. These initial steps are followed by an enormous complex of physical and chemical events which follow inevitably from the initial conditions set up by the gene products, and these events follow in a sequence which is determined by the initial gene-produced proteins, the chemicals they in turn produce, and the physical forces that come into play—diffusion, adhesion, electrical fields, and others that put the chemicals in a pattern. If these reactions and forces move in the same way generation after generation, further genes will seep into the system to fix these processes and make them more stable (the Baldwin effect). There are clearly some internal clock mechanisms (and these are very likely varied and numerous) which see to it that all this complex of events is appropriately timed so that the period of developmental size increase proceeds in an orderly fashion. It is in this way that a consistent pattern is achieved in successive generations.

As I will show in later chapters, it is possible to have an even greater distancing between the initial gene instructions and the final result. This is evident in the process of behavior and especially in learning. An animal with a brain can do things which are initially quite dependent upon the genes that act during development, but once the brain has developed, the activities of the animal may go well beyond anything that could have been anticipated from the gene instructions. But before these matters are examined I want to show that the part of the life cycle we call development itself can evolve.

Chapter 5

BECOMING LARGER DURING
EVOLUTION

THE GREAT lesson that comes from thinking of organisms as life cycles is that it is the life cycle, not just the adult, that evolves. In particular, it is the building period of the life cycle—the period of development—that is altered over time by natural selection. It is obvious that the only way to change the characters of an adult is to change its development.

Let me now examine the way natural selection operates on development at a deeper level. I have already made clear that selection acts on the system of inheritance, namely the genes; they are the ultimate objects of selection, although selection reaches them by selecting the individual organisms that contain them. Then how do we account for the fact that development consists only partially of gene signals, and that on top of those signals there is a superstructure of chemical reactions and physical forces that constitutes the whole of development? The answer must be that all those secondary reactions and processes require genes to start them off in the right direction, and to continue to monitor them as development proceeds. In other words, genes serve as the control elements for the principal chemical and physical processes of development; without the genes calling the shots there would be chaos. The genes see to it that the development of a nematode worm is different from that of a fruit fly, or that of an oak tree, and it is the genes that see to it that development is consistent from one life cycle to the next for each of these organisms.

Because there is so much more to development than the expression of genes, some biologists object to saying that development is determined by a genetic, or DNA, program. This does not bother me so long as one appreciates that the genes give the controlling signals and that the majority of the processes of development do not directly involve the immediate proteins produced by the genes. That many of the events in development, especially in a large, complex organism, may be very far removed from the initial signals from the DNA does not mean that the signals are not important. They are vital; they are the way the initial conditions are set for all that follows. Were this not so, the whole mechanism of evolution by natural selection would be impossible. The genes are like an architect and his foreman in charge of a building construction crew; without them there would be no supervision of the process of putting the building together. It is true that some steps of construction follow one another automatically, but if there were no way of having an ordered program there could be serious consequences—one could not pour the concrete for the basement before building the forms to hold it. Those who don't believe that development is a program set in the genes emphasize that there is so much more to development than just the DNA instructions. As we have seen, in this they are plainly correct, but it is important not to be carried away: both the genes and the construction materials and their properties are essential to development. There could be no development, no life cycle, no evolution if any of these were missing.

There are, nevertheless, problems, especially when one considers the development of large, complex organisms. If the product of any one gene will have many secondary and tertiary effects (and beyond), then it could be very difficult to change that gene by mutation, for any tinkering with its message is bound to affect all the reactions and processes that follow, and for that reason it would be very difficult for any change to occur without detrimental consequences. It appears that genes for different parts of development, such as for different organs in a vertebrate, can act independently from the genes involved in some other part. In this way it is possible to have a change in one part or organ without having dire effects on the whole organism. I have previously called this system of isolating genes into units that can behave relatively independently of one another "gene nets."

A good example of how one gene net can change without affecting the other gene nets is seen in the timing of when certain organs appear during development. For instance, in some amphibians the ovaries and the testes will mature in the larva rather than in the adult, so that in certain salamanders the sexually mature individuals will still possess larval gills. The development of each organ is governed by a separate gene net, and as a result the timing of the appearance of the organs is independent one from another. Such shifts in timing are called "heterochrony." In this case the timing of the genes affecting the development of the sex glands are quite separate from those affecting the timing of metamorphosis, that is, the transition from gill bearing (water breathing) to lung bearing (air breathing) individuals.

Without separating groups of genes into gene nets, it would be difficult to build an animal or a plant of any appreciable complexity. The same can be said of constructing a building. The foreman could decide to have the electrician put in the electrical system either before or after the plumber installs the plumbing. The two systems can be built in either sequence and have no harmful effects on the ultimate building. On the other hand, there is a limit: the roof cannot be built before the cellar any more than the eyes of a vertebrate embryo can be built before the main axis of the embryo has been established. Gene nets have their limits—only some parts of the developing embryo are dissociable and can undergo changes in their timing of appearance or heterochrony. This is what one would expect of any process that unfolds sequentially. In development, all those processes that are launched by the expression of genes and give rise to a whole chain of chemical reactions and sequences of physical steps are bound to have considerable inflexibility for change. But wherever it is possible to tinker by mutation-induced changes without wrecking the whole developing organism, such kinds of changes will occur during the course of evolution, and they will be greatly helped by having a system of gene nets. For that reason, gene nets themselves will be encouraged and favored by natural selection.

.   .   .   .   .

One of the great sins in evolutionary biology is to subscribe to the idea of progress in evolution. However, this is precisely the sin I am about to commit—to claim that evolution does indeed lead to progress. Let me show why both sides of the argument are right.

The reasons for claiming that the idea of progress is unacceptable for any theory of evolution go back at least to Aristotle. His ancient idea, which did not embody any notion of evolution, was that all animals and plants were clearly not equally complex and that at one end of the scale one had plants and lowly worms and at the other end human beings. This so-called scale of nature was considered to be a progression towards perfection, that is, mankind. It hardly needs saying that this scheme will not do today, no matter how pleased we may be with ourselves.

The more modern form of this objectionable progression was one in which evolution seemed to be directional. If one looks at the fossil record, there clearly are trends which seem to show an increase in size and complexity. These facts are not controversial today, but we disagree with their early interpretation. In the latter part of the last century, the paleontologists, who were the main evolutionary biologists, had discovered many of these trends among their fossils. It was suggested that there was some sort of force in nature which led evolution, a force which was given the grand name "orthogenesis" to indicate these straight lines of forward evolution. There is no one today (or perhaps I should say, I know of no one) who subscribes to this mystical idea that evolution is goal oriented and that there are directional trends in evolution. This is the kind of progress all biologists reject, so much so that it has given the word "progress" a bad name. The reason for the rejection is very simple. Such orthogenic trends can easily be explained by natural selection, so why resort to an explanation which is no explanation at all? Orthogenesis assumes an inner drive, a vital force that pushes evolution in a particular direction. There is neither any evidence for such a force nor any attempt at an explanation as to what the force might be. It is pure mysticism, and for that reason alone it deserves the total rejection it gets today. No wonder "progress" is considered by some an outlawed word.

Now for the other side of the story: it is the same argument I used in considering the evolution of multicellularity. If one looks at the fossil record over time, it is clear that the first organisms on earth were single-cell bacteria of some sort. Since then there has been a steady increase in the upper limit of the size and complexity of organisms, so today we have woody plants the size of giant sequoias, and mammals the size of blue whales. Who can say this increase is not progress of some sort? The bacteria and all the middle-sized organisms exist today too, so there has not been a progressive supplanting of one size level by another. Rather, it has simply been a progressive increase in the upper size limit. I will now give more of the evidence to show that this is so, and then try to convince the reader that this expansion of the upper size limit has occurred by Darwinian natural selection. It is one more example of the extraordinary importance of Darwin's idea.

. . . . .

I cannot resist an aside which has to do with my permanent fascination with Charles Darwin, in which I am not alone. A short while ago I made an interesting discovery. I was rereading what is possibly my favorite Victorian novel, Mrs. Gaskell's *Wives and Daughters*, and it suddenly dawned on me that Mrs. Gaskell had made Darwin the hero of her book. As a young man the hero goes on a long voyage to distant lands and comes back mature, bronzed, and suddenly the center of attention of the great naturalists and scientists of his time. Even his personality fits my conception of the sort of person Darwin might have been. I went to our English department and found mild interest in my hypothesis, but no one knew if this had been pointed out either at the time the book was published or later. Some time later I mentioned to a young English scholar who was married to a biologist that I was a fan of Mrs. Gaskell, and she asked if I knew that Mrs. Gaskell was Darwin's cousin and that the two families had often visited one another! This made me doubly convinced that my hero and Mrs. Gaskell's were one and the same.

. . . . .

Over the years I have been interested in the idea that the maximum size of all the organisms on the surface of the earth at any particular time is continuously increasing. With the help of a number of kind paleontologist friends, I have been able to draw a curve showing this maximum size increase over geological time (fig. 14). The earliest known fossils are of bacteria-like organisms found in rocks that are 3.4 billion years old, a little over a billion years after the earth itself was born. No other fossils have been found in these early rocks. The next step was the appearance of the larger, multicellular cyano-bacteria which we have already discussed, for they were among the early pioneers in multicellularity as a means of becoming larger. Eukaryotes did not appear until about 2.1 billion years ago, as small algal filaments. A long period up to the precambrian (about 600 million years ago) has a very poor fossil record. But then one finds a burst of fossil remains, the most remarkable of which are the famous Burgess shale deposits in Canada and similar deposits in China and elsewhere. These invertebrate fossils have attracted such attention because the impressions of their soft parts have been preserved. From them we can plainly see that there existed many strange animals at that time that bear no obvious similarity to the animals of today. Among these early organisms were the trilobites, segmented, arthropodlike animals that became successful and very common. The armor of their outside shells has been well preserved in the rocks.

Favorable conditions for preservation and the increased number of animals with some kind of skeleton indicate that

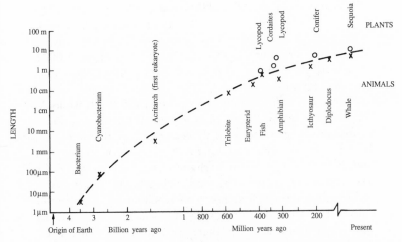

Fig. 14 A graph showing a rough estimate of the maximum sizes of organisms at different periods of life on earth. Note that the length (or height) of the organisms and the time are on logarithmic scales. (From Bonner, *The Evolution of Complexity,* Princeton University Press, 1988.)

during the most recent 500 to 600 million years, there have been more or less continuous deposits of animals of many different species, and at least for the recent 400 million years, there has been a good record of plant evolution as well (fig. 14). From this rich fossil material we can find the largest known plant or animal from any one epoch. Though the graph shows that the largest plants seem to be slightly bigger than the largest animals, this is somewhat misleading because *length* is used as the measure—the only reliable one for fossils. If we could *weigh* the actual organisms, animals would prove to be at least as large or larger than plants.

Figure 14 also shows that the largest plant that ever existed, the giant sequoia, is still alive today. Similarly, the largest animal, the blue whale, still swims in our oceans, although in greatly reduced numbers.

The range of size in this figure is certainly enormous—about twenty orders of magnitude if one uses weight as a measure. Small bacteria weigh roughly 10^{-12} grams, while a blue whale weighs about 10^8 grams, that is, there is a range from

0.000000000001 to 100,000,000 grams. If we use length instead of weight as a measure, the difference does not seem quite so large, but it is still staggering. A bacterium may be 10^{-4} centimeters long and a whale over 10^3 centimeters (.0001 to over 1000 centimeters), which is a difference of seven orders of magnitude.

Remember that this range of sizes exists today. If we look, for instance, in the forests of North America, we will find a continuous range of sizes. At the lower end are the omnipresent bacteria. Amoebae and other protozoa feed on the bacteria and they in turn are eaten by slightly larger nematode worms and other invertebrates. The food chain includes invertebrates of increasing size, such as insects and their larvae, and these in turn will be food for shrews, mice, and small birds, and at the upper end of the size scale there are squirrels, rabbits, foxes, bears, deer, and moose. One can show a similar array of sizes for plants, from algae and fungi to mosses and small grasses, to shrubs and finally large trees. Lakes and oceans have the same gamut of sizes of animals and plants, the largest being whales and kelp (the huge brown algae that can stretch well over a hundred feet in length), respectively.

．　．　．　．　．

I could describe with fascinating detail the largest flower, the smallest insect, the largest invertebrate, and the smallest vertebrate, but it is not my purpose here to display a sideshow on the wonders of nature. Rather, I want to demonstrate that the size of an organism has a profound effect on its shape, on how it moves, how it eats, how it performs all its physiological functions. Size by itself is a potent factor in determining what animals and plants look like and how they work, and it is of interest to see how that can be. It is an inquiry into the mechanical engineering of organisms and the role of size in their design.

This subject is hardly new—in fact, it is a venerable one that dates back to Galileo and other Renaissance scholars. Recently, however, it has, in various forms, gone through a new renaissance, and I will concentrate here on these revolutionary developments. The earlier work is elegantly summarized in D'Arcy Thompson's great book, *On Growth and Form*.

First I want to say a few words about D'Arcy Thompson and his work. He was a towering man, both physically and intellectually, and a professor of zoology at St. Andrews University in Scotland for many years that spanned from the end of the last century well into this one. Besides being a biologist, he was a distinguished mathematician and an even more distinguished classicist. His translation of Aristotle's *History of Animals* is still a standard one. He also wrote a *Glossary of Greek Birds* and another on Greek fishes wherein he cites every reference in classical Greek literature to these animals, identifies the species, and describes what we know of their modern zoology. His great achievement was *On Growth and Form*, an extraordinary fusion of science and history and their bearing on what we understand of biological form. More than this, it is written with an elegance that we have lost today. I will give one example, to persuade my reader to read his book—a passage from its philosophical introduction. One may not agree entirely with his position, but one certainly has to admit that he has stated his case wonderfully well.

> How far even then mathematics will suffice to describe, and physics to explain, the fabric of the body, no man can foresee. It may be that all the laws of energy, and all the properties of matter, and all the chemistry of all the colloids are as powerless to explain the body as they are impotent to comprehend the soul. For my part, I think it is not so. Of how it is that the soul informs the body, physical science teaches me nothing; and that living matter influences and is influenced by mind is a mystery without a clue. Consciousness is not explained to my comprehension by all the nerve-paths and neurones of the physiologist; nor do I ask of physics how goodness shines in one man's face, and evil betrays itself in another. But of the construction and growth and working of the body, as of all else that is of the earth earthy, physical science is, in my humble opinion, our only teacher and guide.

I will begin by summarizing (in my own words) the initial main points in his chapter "On Magnitude," and then go on to current considerations.

The first principle is that the surface of an organism goes up as the square of the linear dimensions (l^2), and the volume by the cube (l^3). If one were to imagine two spherical organisms, one a millimeter in diameter and the other ten times thicker, that is, a centimeter in diameter, both would have the same formulas for the ratio of their linear dimensions (radius) to their volume or their surface:

$$volume = 4/3 \ \pi r^3 \ (v \propto l^3)$$
$$surface \ area = 4 \ \pi r^2 \ (s \propto l^2)$$

As we go from the small sphere to the larger one, the volume rises as the cube of the linear dimensions, while the surface area rises only as the square. This means that the bigger organism will have a larger proportion of volume to surface than the smaller one. This simple fact has enormous and far-reaching consequences.

Imagine that our two organisms need oxygen in order to live, and that the oxygen is consumed in relation to the volume of the sphere. Oxygen can get into our hypothetical animal only by diffusing through the surface. In the small sphere, sufficient oxygen can easily diffuse inward to satisfy the modest oxygen needs of the small organism, but in the larger one there is too little surface to accommodate the comparatively much larger volume (l^3 compared to l^2); without some other help, the larger organism would quickly perish. The problem could be solved by reducing the internal metabolism, and therefore the demand for oxygen; but this would be an inadequate method of rescue, largely because a sphere ten centimeters in diameter is so thick that it would take a very long time for the oxygen to diffuse into the central part of the sphere—too long to prevent the internal part of the sphere from dying.

Living organisms have found a number of ingenious ways of solving the problem. One is that with increased size the surface increases disproportionately, so that there is no longer a neat sphere but a structure with a skin that forms great folds into the interior. There not only is more surface for the oxygen to penetrate, but the folds are such that no part of the interior is far removed from the oxygen-supplying surface. Our own

bodies have many key surfaces that are required for the diffusion of substances. Oxygen diffuses to our blood in the lungs,
and because the lungs are made up of millions of small com-
partments (or alveoli) their surface is greatly increased and,
therefore, so is our ability to pick up oxygen. Furthermore, the
blood stream moves continuously due to the pumping of the
heart, picking up the oxygen in the lungs and depositing it in
the internal tissue where it is needed. Without these devices we
could not be much more than a millimeter tall. Conversely, a
single-cell protozoan can exist happily as a small drop of sim-
plified protoplasm only because it is so small.

We absorb other things besides oxygen through surfaces.
For instance, the food we take in is mostly broken down into
small molecules such as sugars and amino acids, which are absorbed through the wall of our intestine. Because of the sur-
face-volume principle, one would predict that the larger the
animal, the longer the intestine and the more numerous the
internal convolutions so that the surface could catch up with
the increase in volume. This is precisely what happens. The
smallest nematode worms have a gut that is a simple, straight
tube, while we have an enormously long intestine with a fantastic number of protuberances that vastly increase the surface
to compensate for our great increase in volume.

Turning to the question of strength, we find that surface is
not the only part of a living organism that varies as the square
of the linear dimensions (l^2). The strength of any part of a
body varies with the cross-section area (l^2) of that part. For
example, if a man has a very thick biceps we correctly assume
he is strong, and a spindly arm indicates relative weakness. The
same principle applies to bone—the thick leg bone of an ox is
impossible to break, while that of a chicken can easily be
snapped in two.

However, the weight of an animal or an arm or a bone varies
with the volume and therefore as the cube of the linear dimensions (l^3). This means that as organisms become larger, their
weight increases far more rapidly than their strength. If we
hold a pencil by placing its ends on a finger of each hand, we
note that it is strong enough to span the distance without sag-

ging. Now, suppose we enlarged the pencil so that it was big enough to span the Hudson River: we would immediately discover that it would collapse under its own weight, for its weight has gone up much faster than its strength. This explains why the George Washington Bridge does not look like a pencil!

In fact, everything about bridge building is so arranged that strength can be maintained and weight decreased. This is the reason for the overall design, including the thin cables, the lacework of trusses, all of which are clever tricks of the engineer to keep the strength-weight ratio within safe bounds. Hollow cylinders may be as strong as solid rods, but they weigh much less, a principle we see well illustrated in bamboo. In construction the I-beam steel girder gives the strength of a great square rod of steel, but has only a fraction of its weight.

To give more biological examples, fruits such as apples and cherries are supported by stems whose cross-section areas must be sufficiently thick to give the strength necessary to support the fruit. This is why melons and pumpkins lie on the ground and don't hang from trees. A stem thick enough to hold a large watermelon in the air would probably equal the diameter of the watermelon itself—not to mention the problems of the trunk of the tree which would have to bear it.

As I have said, in the case of bridges and other feats of engineering, the main way to solve the strength-weight problem is by reducing the weight. This solution is not open to animals and plants of increasing size, for their weight must necessarily increase as they become larger. Therefore, organisms have devised ways to increase their strength to support the increase in weight. To give a good example, which comes straight from D'Arcy Thompson, one can determine the percentage of weight of the bones of different-sized animals in relation to their total weight and show that with increased size the percentage of the weight that is bone increases. In a mouse or a wren, bone is 8 percent; in a dog or a goose, 13 to 14 percent; in man, 17 to 18 percent. It is a pity that Thompson does not have information on an elephant as well.

Obviously as weight increases there is a disproportionately large increase in the thickness of the support structures, such as bones, which provide the strength. Our attempt to understand this relation has made the study of size in organisms such a lively and interesting subject today. We presume that the increase in strength necessary to accommodate the increase in weight is the result of natural selection, for only animals and plants that remain competitive with increase in size will survive and be successful in producing offspring. The question is, what property of the organisms is selected for? Selection works on the simple principle of trial and error, so we must follow up with the next question: Is there some way of predicting what is the best strength-weight construction so that it will inevitably be the winning design in the process of natural selection? One of the most interesting recent hypotheses, for which there is much support, has been proposed by Thomas McMahon. He suggests that the elastic properties of organisms are crucial for their optimal design. In other words, the proportions of animals and plants of different sizes should fit the prediction that they keep the same elastic properties and in this way avoid buckling due to an increase in weight.

It has been recognized for centuries that the bigger trees are disproportionately thick. This is a common observation we have all made: a young sapling will be thin and willowy, but a great oak will be exceedingly thick and massive. Our largest tree, the giant sequoia, looks ponderous in its proportions compared to a smaller pine tree.

On purely theoretical grounds McMahon predicted that if elastic similarity is maintained between trees of different sizes, the height of the tree should be proportional to the diameter of the trunk to the two-thirds power ($h \propto d^{2/3}$). This relation can conveniently be plotted on log-log graph paper, giving a straight line. This is because $h \propto d^{2/3}$ is the same as $\log h \propto 2/3 \log d$, which is a straight line relation. McMahon found a book with the rather unlikely title of *The Social Register of Big Trees*, and he plotted all these trees on logarithmic graph paper (fig. 15). They came out (with considerable scatter) very close

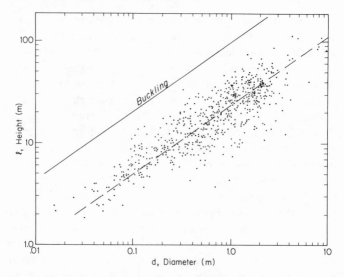

Fig. 15 Diameter of the base of trees plotted against their overall height on a logarithmic graph. The dots are 576 record specimen trees, representing what is believed to be the tallest and the broadest of each species found in the United States. (From T. A. McMahon and R. E. Kronauer, *J. Theor. Biol.* 59 [1976]:443.)

to the predicted slope of two-thirds. Furthermore, he could calculate at what point the trees would reach their elastic limit and buckle, and it was possible to show all were well on the safe side of this limit of their strength. Trees bend in the wind, and they sway back and forth like an inverted pendulum in a gale. They manage remarkably well, although they may reach and even exceed their limits in a hurricane or a tornado.

The message that elastic similarity arises when natural selection favors equal durability for different-sized trees is driven home if we compare large trees with minute, treelike fruiting bodies that stick up into the air and are also at the mercy of the elements. As I pointed out earlier, these are found among many minuscule fungi such as molds and mildews and among slime molds. The latter are a useful example because they vary greatly in their size, and one can easily compare their proportions as was done with trees. But they differ strikingly from trees because with increased size their proportions do not change; they do not become disproportionately thick. In other words, the relation of their diameter to their height remains constant, regardless of size (h $\propto$ d). This is exactly what one would have predicted with these miniature "trees" that range around a few millimeters in height. Their weight is negligible, and therefore a small one, as compared to a large one, is equally unaffected by gravity; there is no need to become disproportionately thick. Natural selection would not have pushed the fruiting body's proportions in this direction.

Animals move, and the rate at which they move is in general directly proportional to their length (velocity $\propto$ length). This is true for both swimming and running animals. The relation is slightly different for flying animals (v $\propto$ $l^{1/3}$), but it remains true that a wasp flies faster than a small fruit fly, and a swan flies faster than a sparrow (fig. 16). In the case of running animals, the smallest are mites about a millimeter long, and the largest would be mammals of over a meter long. Again, mites are slower than larger ants, squirrels are slower than foxes, and a cheetah is the fastest of them all. The relation breaks down somewhat for the very largest mammals, for elephants are not faster than cheetahs or even horses. The relation for swimmers also breaks down, for a large tuna fish is faster than a whale. But among fishes the relation holds true; even the baby fish are slower than adults of the same species, and as they grow, their speed increases linearly with their length. The relation holds right down to ciliate protozoa down to the smallest motile bacteria—an enormous range of sizes spanning over six orders of magnitude.

I might add, parenthetically, that I had a most interesting time collecting all the data to show this relation between size and speed. I could find nothing on mites at all, so a great expert on these beasts kindly ran some mite races with different species and sent me the results; the rest came from combing the literature. I discovered, for instance, that the speed of a few species of ants had been determined and published by the distinguished astronomer, Harlow Shapley. He found that the running rate of ants on Mount Wilson near the observatory

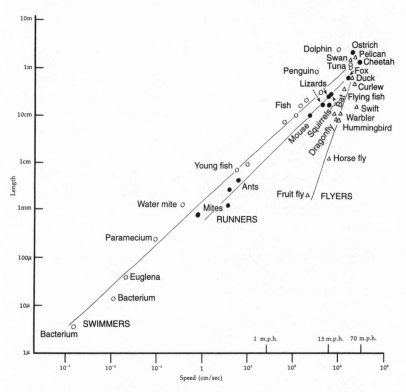

Fig. 16 The maximum swimming, running, and flying speeds of organisms of different size. The length of the animal is plotted against the velocity, both on logarithmic scales. The data have been selected for a diversity of types of organisms and for the most rapid examples within each group. (From Bonner, *Size and Cycle*, Princeton University Press, 1965.)

varied in a very precise way with the temperature, so that he could actually determine the temperature of the day by clocking the ants. This paper came as a surprise to me, for I suddenly realized that astronomers have relatively little to do in the daytime.

The proportional relation between size and speed is remarkable when one considers that we are dealing with a number of different modes of locomotion, especially when we compare running with swimming. Even swimming, as we shall see, is performed in radically different ways depending on size. All of running involves legs, but the numbers vary: mites have eight; insects, six; most mammals, four; birds and humans, two, along with kangaroos, but they hop instead of run. Most of the recent work has been done on quadrupeds, which we will now consider.

In an important paper which appeared in 1950, A. V. Hill, the well-known muscle physiologist (who received the Nobel Prize for his work) suggested that all four-legged mammals should run at about the same speed, regardless of their size. His argument, in a nutshell, was that if one animal were ten times larger in height and length than another, then its stride would be ten times longer, but its muscles, which also would be larger by a factor of ten, would move ten times more slowly. Therefore, a mouse would take ten rapid steps in the same time it took a dog (ten times larger) to make one step, and as a result they would exactly tie in a race. He gathered what scant data there were at the time on running speeds among different mammals, and as one would expect they did not come out to be exactly the same. He suggested that the variation in speed came from the difficulty of measuring maximum speed in wild animals and, besides, each animal was not similarly constructed, for if one compared, for instance, a cat with a dog, they move in a very different fashion. Both of these difficulties are indeed true, and we now not only have better ways of measuring comparative speeds, but we also recognize that quadrupeds of different size are not geometrically similar at all. The larger the animal, the thicker the limbs, as is the case with trees. We can also see this if we compare a graceful gazelle and its slender legs to an elephant and its four huge stumps.

The first problem, that of obtaining better figures for running speed, was solved by C. R. Taylor and his associates who ran all sorts of animals on treadmills. They found that the maximum speed was an elusive figure to obtain, and a much more reliable one was the speed at which the animal switched from a trot to a gallop. If they compared this measure for animals that vary greatly in size, like a rat, dog, and a horse, they found that the speed of the transition from trot to gallop was not constant, but increased with the size of the animal. Furthermore, they showed that the speed went up by one quarter power of the weight (velocity $\propto$ weight$^{1/4}$), which is precisely the relation that would be predicted from McMahon's elastic similarity hypothesis.

The role of elasticity in running is far greater than one might expect at first. All the muscles, and especially the tendons in a body, are elastic and behave like rubber bands. This means that if a tendon is stretched, it is storing elastic energy, for like a rubber band it will contract of its own accord after each stretching, so that much of the energy required for the stretch is not lost but saved for the contraction. To illustrate the truth of this statement, the Taylor-McMahon group had a surgeon implant two steel balls in the tendon of a cat, and then ran the cat on a treadmill, all the while taking a motion picture of it with an X-ray camera. One can plainly see the two balls in each stride cycle stretch far apart from one another only to come rushing back together. When I saw this film it made quite an impression on me: not only did I see a vigorously running cat skeleton, but because there was something wrong with the way the film was loaded, the ghostly skeleton ran along the ceiling instead of the floor.

Another impressive way in which the group showed the significant role of storing elastic energy was by putting a kangaroo on a treadmill. They hooked it up with a respiratory mask so they could measure how much oxygen the animal was consuming, that is, how much energy it was burning up. If the treadmill was going fast enough so that the kangaroo could hop in a steady lope, the amount of oxygen it burned was reasonably low and did not rise very much for considerably faster speeds. However, when the treadmill was run very slowly, the kangaroo would intermittently give one hop and then pause before the next hop. This slow maneuver requires much more energy than the faster running because each hop is a separate event, and there could be no storage of elastic energy as with the rhythmic continuous movement during normal running.

The elasticity is stored not only in the legs but in the whole body. One can see this dramatically when watching a cheetah run after a gazelle in one of those grand nature films. With each stride the whole spine and tail of the graceful creature undulates. This wavelike motion along the back is elastic, and each stretch goes on to contract, thereby greatly enhancing the efficiency of seventy-mile-per-hour sprints. Some animals, such as horses, have a rigid back (it would be very hard to ride a cheetah!), so that clearly the spine is not storing elastic energy. However, there is evidence that the ribs of a horse, which expand and contract so each stride is synchronized with inhaling and exhaling a breath from the lungs, in this way not only store the elastic energy needed for running, but also facilitate breathing—all in one economical motion.

So far we have considered only the relation between size and speed in large organisms, and we have confined our examples to running quadrupeds. We could have made similar arguments for swimming and even for flying, but it is time to move to the problems of minute, microscopic organisms, and here we are in the realm of swimmers.

Despite the fact that whales, fish, and many kinds of unicellular eukaryotes and bacteria can swim, the problems of the small organisms are totally unlike those of larger ones. Their environment is entirely different, and the only reason for this is that they are so small. To understand this difference we must first discuss a most important dimensionless number called the Reynolds number. Without going into details, the Reynolds number is proportional to the force of inertia of a body moving in a medium, and inversely proportional to the viscosity of the medium. In other words:

$$\text{Reynolds number} \propto \frac{\text{Inertia}}{\text{Viscosity}}$$

Inertia is the resistance offered by any body to a change in motion. Clearly the larger the body, the greater the resistance to a change in speed, so a large fish would have much inertia, but a minuscule bacterium would have very little. The viscosity is the thickness of the medium, which for organisms will be either water or air. Here we are concerned with swimmers, so we will always be dealing with the viscosity of water (except in some of my imaginary examples where I will talk about thick molasses, which has a very high viscosity).

From this we can see that fish live in a world of high Reynolds numbers, while bacteria live in one where the Reynolds numbers are low. It may not be immediately obvious why this makes a big difference. To help the reader understand the problem, I will paraphrase a passage from a talk by the physicist Edward Purcell. He asks us to imagine the conditions of a man swimming at the same Reynolds number as his own sperm. Put him in a swimming pool full of molasses, and then forbid him to move any part of his body faster than one centimeter per minute. Now imagine yourself in that condition: you are in a swimming pool of molasses, and you can only move at the speed of the hands of a clock. Under those ground rules, if you moved a few meters in a couple of weeks, you would qualify as a low Reynolds number swimmer. By using the molasses and by slowing down the rate of movement (and therefore the inertia) of the man, the Reynolds number would come down to the range of unicellular microbes, so his progress would be very slow.

A swimming single cell has another problem. We can lift our arms out of the water (or molasses) at each stroke as we swim, so at least we will go forward, however slowly. But small organisms cannot do this. To see this problem, imagine a man in a row boat who is not allowed to take his oars out of the water—all he can do is push them backwards and forwards. Nevertheless, he can manage for he is in a high Reynolds number situation, and he can pull the oars towards him rapidly and push them back very slowly. Because of its high inertia, the boat will go forward. Now put the boat in Purcell's molasses and allow him to move the oars at a snail's pace. The boat cannot move at all because the difference between the fast pull and slow push will have no effect in this low Reynolds number situation.

Minute organisms solve this problem in two ways. One is the method used by the relatively large eukaryotic single cells such as flagellates and ciliate protozoa. Their cilia are what could be called "flexible oars," for as they push in one direction they will remain erect and stiff, but as they pull back they bend and therefore exert much less force against the water. Because of this stiffness in one direction only, they can avoid the problem of the man rowing in molasses. Similarly, if the man had flexible oars, he could presumably move forward for the same reason. Flexible oars are a simple trick to solve the problem of how to move in a world of small Reynolds numbers.

Bacteria are considerably smaller than unicellular protozoa, and they have much thinner flagella that lack all the complicated fine structure of a protozoan cilium or flagellum. In fact, their flagella are quite incapable of acting as flexible oars. Bacterial flagella do not wave at all but are rotated at the base so that each flagellum corkscrews through the water. This is an even cleverer way of coping with the problem of moving when the Reynolds number is very low. It works on the same principle of the modern corkscrews that remove the cork from the bottle by pushing the cork up the corkscrew as it rotates. The bacterium bores its way through the water just the way a corkscrew "swims" through a cork.

Let me add parenthetically that minute flying insects do low Reynolds number flying. Some minuscule wasps, which for some perverse reason are called fairy flies, have wings that look more like feather dusters than respectable wings. Their inertia is so small that the air must seem like a viscous fluid, and perhaps their feathery wings act like flexible oars as they swim through the air.

•   •   •   •   •

Size increase (or decrease) sets certain physical limitations or conditions on an organism that might be described as mechanical necessities. These are crucial if the organism is to carry out its normal functions: that is, its taking in of energy and processing that energy so that it can be used for its sundry activities. In the case of plants, energy is largely expended in growth and reproduction; but animals, which are active by comparison, use that energy to keep warm, to move, and to keep all the cells of their various organ systems, including the brain and the nerve cells, supplied with energy so they can operate effectively.

Among these mechanical consequences of size increase is a division of labor. I described the beginnings of such a cell differentiation in the evolution of early multicellular organisms from unicellular forms, and with further size increase there has been a corresponding increase in the degree of the division of labor. Without such a parceling out of functions, huge masses of cells, which constitute larger animals or plants, simply could not cope with all the problems of activity needed for existing as efficient, competitive organisms. However, it must be emphasized that this division of labor, which leads to greater efficiency, is of itself selectively advantageous. In other words, natural selection acts both on size increase and division of labor to produce organisms that can effectively see to it that their genes are passed on to the next generation. There is a three-way relation between natural selection, division of labor, and size.

Division of labor is related to the notion of complexity; an increase in one means an increase in the other. Complexity is measured in the number of parts and the interrelation of those parts. The simplest way of applying the term to living organisms is to equate a part with a cell type. In slime molds there are spores and stalk cells; in mammals there are muscle cells, nerve cells, liver cells, gut cells, and innumerable others, and each one of these general categories may be further subdivided into many different specific kinds of cell types. Furthermore, it is possible to show that in a rough way there is a correlation between size and cell types: larger organisms will have more cell types than smaller ones (fig. 17). This does not mean there are no small mammals or insects, for we have shrews and fairy flies. It is quite possible that once a certain group of organisms has evolved, such as insects or mammals, for some species there will be a selection for size increase and for others size decrease. In the latter there is evidence in some invertebrates (such as rotifers) that with size decrease there may be a loss of some cell

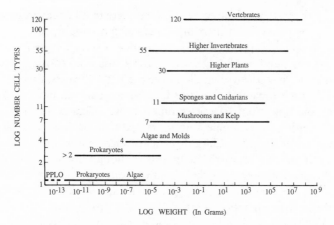

Fig. 17 A graph showing the size ranges (by weight) of groups of organisms containing different numbers of cell types. (From Bonner, *The Evolution of Complexity*, Princeton University Press, 1988.)

types, presumably no longer needed because of the small size. In the case of size increase, in any group of organisms there must be an upper size limit that can be supported with a given number of different cell types, and therefore the only way of achieving an even greater size is to increase the degree of division of labor.

To return to the question of progress, we can now argue that there is both a selection for size increase and an increase in complexity in the sense of a greater division of labor among cell types. Between these two objects of selection, organisms have evolved that are both large and complex enough to co-exist with smaller, simpler forms. As before, this can be called progress only if one refers to the increasing range of sizes and complexities as evolution proceeds: in early earth history there are only very small organisms, but over many millions of years of evolution the range of size and complexities steadily increases until today one has the greatest range ever supported on earth. This kind of progress is acceptable only because it is clear that the mechanism of its evolution can be easily explained by natural selection, and therefore brings us out of the fog of mysticism. I am very conscious here that I am expressing my own views, and there are others who do not agree.

Skeptics cropped up the very moment *On the Origin of Species* was published by Darwin in 1859 and, even though their number has slowly dwindled over the last 150 years, there are still some today. The reasons have often changed, but I think all the objections have a common cause. What is difficult to accept is that so simple a notion as natural selection could account for the astounding diversity in shape and complexity of all the species on earth today. It is as though one were having the great mysteries of the world explained by Beatrix Potter. Many feel that "something" is missing, and for that reason the theory has left them dissatisfied. In the early period after the publication of *Origin*, one of the main objections was that while it was easy to see how selection could have a negative effect and eliminate undesirable characters, it was difficult to see how new structures appeared during the course of evolution, such as wings or complicated eyes in vertebrates, to give two extreme examples. Today we would argue (as did Darwin) that these innovative changes occurred in small steps, while in the nineteenth century the general answer (but one not supported by Darwin) was that besides natural selection there was, as previously mentioned, an inner driving force that produced "progress."

Another great difficulty in the beginning was the lack of understanding of the mechanism of heredity. It meant that many biologists—and Darwin himself is faintly guilty of this sin—resorted to some form of Lamarckism to account for change. They would suggest that change came about through use or disuse, and the change was inherited—in other words, there was an inheritance of acquired characters. Suggestions of Lamarckism are especially evident in discussions of behavior where it was argued that a particular pattern of behavior can ultimately become inherited. There is some truth to this, but as we shall see later, the explanation can be purely Darwinian. In his *Descent of Man* Darwin has an interesting discussion of the inferiority of the intellect of Victorian women, a deficiency he suggests could be corrected by a few generations of more rigorous education. He is right, of course, but it will not become inherited as he seems to suggest; such a change is bound to be entirely cultural and therefore could happen within one generation.

Those who today continue to reject the idea that natural selection alone is adequate to explain evolution are a curious variety of bedfellows. To begin, whole countries seem to adopt an attitude that takes on a sort of national view among the academic community. For instance, Germany, immediately after the publication of the *Origin*, led the pack in approval. To some extent this was due to the influence of a few key biologists. Ernst Haeckel, a tremendous enthusiast and vigorous popularizer, was a staunch supporter of a somewhat muddled version of Darwin's ideas and preached the message with eloquence and ardor all over Germany, and his books were translated into other languages, including English, as well. He had a gift for simplifying science for the layman. This was his strength as well as his weakness, for often his generalizations were too sweeping. I can remember complaining to an older colleague that a theory of Haeckel's on how multicellular animals arose (from an invaginating *Volvox*-like ancestor) had long been discredited. I was greatly shocked in my youthful idealism to hear the professor say, "I know that, but it's so easy to teach." (The irony of this story is that after one hundred years of aggressive neglect, Haeckel's old idea is coming back into favor!) In any event, Haeckel, the colorful, flamboyant propagandist, did much to make Germany pro-Darwin.

A far more important intellectual contribution was made by August Weismann, who, I think, came closer to the stature of Darwin than any other evolutionary biologist of the nineteenth century. He also was a man of great depth. Not only did he understand Darwinism, but he did much to take aspects of the subject some steps further. He pointed out that one could not inherit acquired characters, for docking the tails of lambs or circumcising Jews had never led to the absence of tails or foreskins in the progeny. Furthermore, he pointed out the distinction between the immortal "germ" cells (the egg and the sperm), which produce the next generation, and the "soma," which is the mortal part of the body of a multicellular organism that dies each generation. Note how this foreshadows the views of Dawkins, where the soma parallels Dawkins's "vehicle," and the germ plasm the DNA "replicator." I have come to appreciate Weismann more and more because he was the first to understand how embryonic development was necessarily tightly linked to the problem of evolution. As someone interested in both evolution and development, I think Weismann's ideas are again coming to the fore and are being appreciated anew. In his day he was greatly respected and his influence stimulated the acceptance of natural selection in Germany. I do not mean to imply that everyone in Germany was converted, but certainly a large majority was. There are some interesting exceptions; for instance, Hans Spemann, who discovered embryonic induction, was very much against the notion that selection is the driving force of evolution.

In America and Russia the acceptance was more slow, and it was not until the early part of this century, with the help of eloquent advocates such as E. G. Conklin, that the pro-Darwin forces became dominant. In Russia, Darwin was always held in high esteem. It is an interesting fact that while Russian biologists were quite satisfied with natural selection, the more philosophical Marxists of recent years have been less enthusiastic. There was the shocking period of Lysenko, when Lamarckism became the official state doctrine, but that was a political rather than a scientific lapse.

An exception to the acceptance of Darwin's ideas was, and still is, a strong force in France. I do not completely understand why, but countries that have their own language and their own vigorous intellectual community can generate views that differ from those of their close neighbors. It might be simply the lingering influence of Lamarck, who is still rightfully admired, but that cannot be the real answer. It is true that in politics France has often shown its independence and has cultivated its home-grown positions with little worry that they may be different from those of others. It is the spirit exemplified by the independent mind of Charles de Gaulle.

There are many first-rate biologists in France today who do not believe that natural selection is sufficient to explain evolution. A good example was the French-Swiss child psychologist Jean Piaget. He was a dedicated Lamarckian and in his early years made some interesting observations on lake snails that appeared to show the inheritance of acquired characters. A number of other distinguished biologists have written papers and books in which they stress the inadequacy of natural selection. What they substitute is very difficult for me to follow except when elements of Lamarckism protrude. The feeling of the inadequacy of Darwin's ideas is all too evident, but I become impatient trying to understand how they think evolution may be explained; I get lost in a miasma of words. I cannot avoid having dark suspicions that, as in orthogenesis, they favor some unidentified force which is the cause of evolutionary change. I vividly remember going to a lecture in Paris in the 1950s given by a venerable and very distinguished zoologist on the inadequacies of Darwinism. In the discussion that followed he was vigorously attacked by a Catholic priest who was all in favor of natural selection, which seemed to me at the time, and still seems to me, an interesting bit of irony.

THE ADULT PERIOD

We have now arrived at the stage of the life cycle that can be considered the adult period. An adult is roughly defined as an individual that has reached the period in its life cycle when it can reproduce. This definition is best suited to large, sexual organisms; the catagory is not so clear in some primitive organisms with asexual cycles, but that is a detail. The important point is that those adults, in order to survive and successfully reproduce, have evolved all sorts of interesting ways to manage within their environment. They have clever ways of finding food, shelter, and mates, avoiding predators, caring for their young, and adapting in many other ways. All these remarkable skills are the result of natural selection, for they contribute to success in reproduction. Success in producing offspring that in turn survive results in a continuous succession of generations.

The most spectacular adaptations of this sort are found in animals (the adaptations of adult plants are relatively simple, although they serve their purposes well). Active animals become so remarkable in their adult lives mostly because they have a nervous system and are capable of behavior. In the next three chapters we will learn about the different stages in the evolution of behavior. First I will discuss simple awareness, then the formation of animal societies, and finally the acquisition of culture.

Chapter 6

BECOMING AWARE

WITH respect to the increase in the range of sizes of organisms during the course of evolution, one particular part of an animal has had an enormous influence. It is the brain, for the brain is responsible for behavior, which, as we will see, can do things that affect evolution in a unique way. At first it would appear strange that this major invention is something confined to animals, for certainly plants are without brains. On the other hand, brains did not burst forth in one great leap: their makings are found in properties of all organisms. The range in the ability to "behave," from the simplest to the most complex way, seems to be another property of organisms that has increased during the course of evolution. In its essence it is no different from any other division of labor, but its consequences for evolution far exceed any other of the special ways an organism handles a particular function or labor. These consequences have had a major effect on the course of evolution.

The most primitive kind of behavior one can imagine is where an organism responds to something in its environment. It might be movement away from or towards light or some other property of the environment, such as moving towards food or away from some toxic chemical. These are properties we find in all organisms, from the simplest single cells to complicated higher plants or animals. It is in this way that organisms show an awareness of their environment. However, there are different levels of awareness, and awareness of the immediate surroundings (and the appropriate response) is at the bottom of the scale. At the other end is self-consciousness, which we think of as a particularly human trait. We also call it consciousness, or the ability to be aware of ourselves so we do not act like robots. There has been great interest recently, largely through the work of Donald Griffin, in the idea that other animals have some degree of self-awareness and that therefore human beings are not unique in this respect; they simply have a higher degree of consciousness than nonhuman animals.

.   .   .   .   .

I shall begin with the most primitive displays of awareness rather than the most advanced. Of particular interest are examples where a simple organism responds not merely to the environment, but to members of its own species. These represent the first instances in which organisms are able to communicate with one another; the first step towards some sort of social integration. Because this is a subject that has absorbed me greatly over many years, I will use the slime molds as an example.

The feeding amoebae are known to communicate with one another in a form that is perhaps the most primitive of all kinds of communication between cells. It is possible to show that the single slime mold cells repel one another. There is good evidence that they give off some substance that causes cells around them to move away, a process known as negative chemotaxis. Presumably this might be a help for more efficient grazing of bacteria by the solitary amoebae, or perhaps it is simply a selfish defense of a food supply. After the bacterial food supply is consumed, the amoebae aggregate to central collection points. This is again because of a chemical substance they give off, but instead of being a repellant, it is an attractant that causes the unicellular amoebae to come together, in groups. In the process of coming together, one aggregate forms a single multicellular organism; the communication between organisms (the separate cells) ceases when cell contact is established, and now all communications between the cells in the cell mass are internal, developmental, cell-cell signals. There is increasing evidence of many such internal signals which regulate all aspects of development, including the ratio of spores to stalk cells.

This is the background to the story I want to tell. When I

began working as a research student of William H. Weston, he would ask me more than once, "Why do slime molds, and all small fungi with fruiting bodies, rise straight up into the air at right angles to the substratum, no matter what the orientation of the substratum?" One can place a culture dish so that its surface is vertical, like that of a wall, yet each fruiting body juts out at right angles from the surface. (These experiments must be done in the dark or in uniform light, since the rising fruiting bodies move towards light.) I have worked on and off on this problem for the last fifty years—in fact, I am still working on it now.

My first experiment that bears on the problem was one that, at the time, I had no idea was relevant. I cut a migrating slug up into three pieces and noticed that the front piece, when it fruited, leaned forwards; the hind fragment leaned backwards, and the fruiting body of the middle piece was somewhere in between (fig. 18). Since this was done when I was an undergraduate, I am not surprised that my interpretation of it was rather odd: I thought the front end fruited while leaning forwards because the fast cells were in the front, and those in the hind end were slow, causing it to drag its heels backwards when it fruited.

The folly of this interpretation became obvious when, with the help of two seniors doing their honors thesis with me, we moved the three pieces around, as in the shell game. No matter where the piece came from in the original slug, it always leaned away from the other cell masses. There seemed to be a repulsion between the rising fruiting bodies.

Fig. 18 A large slime mold slug that has been cut into three pieces (the anterior end is at the right), and as the three fragments rise up into the air to fruit they lean away from one another. (Drawing by K. Zachariah.)

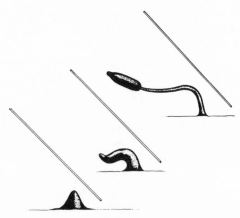

Fig. 19 A slime mold fruiting body rises under a small slip of glass, and is deflected. (Drawing by K. Zachariah.)

After the seniors were graduated, my assistant and I decided to look further into the matter to see what might be the mechanism of repulsion. We were able to show in a series of experiments that the cell masses must be giving off a volatile substance, a gas that was repelling neighboring cell masses. Some of the experiments were quite far out. For instance, we decided to see what would happen if we placed the rising fruiting body in an air current. To do this I went to the Department of Aeronautical Engineering, and with their help built the world's smallest wind tunnel. When moist air was passed over a rising cell mass, it leaned into the wind, suggesting that the wind blew the repellant gas behind the rising mass, in that way causing it to move upwind, away from the repellant. We also placed two incipient fruiting bodies close together. They would lean away from each other as they rose, but if the primordia were not so close, their angles of leaning apart would be smaller. If a cell mass was put in a cavelike crack, it would rise so that it was equidistant from the roof and the floor of the minicave (fig. 19). If the mass was placed on the edge of a cliff of agar, it would jut out, exactly bisecting the distance between the top surface of the cliff and the vertical wall below (fig. 20). And of

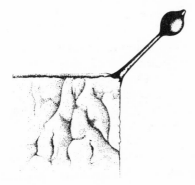

Fig. 20 A fruiting body rises from the edge of the agar
cliff and exactly bisects the angles between the two
surfaces. (Drawing by K. Zachariah.)

course, as I knew from my old teacher, they rose at perfect
right angles from a flat agar surface.

The really crucial experiment was to add a piece of activated
charcoal which has the property of rather indiscriminately absorbing and binding any volatile substance. If this was done
the fruiting bodies near it would make a nose dive right into
the charcoal (fig. 21). This was because the charcoal removed
the repellant gas on its side of the rising cell mass.

Since the whole process of fruiting as well as the movement
of the migrating slug involves the movement of the individual
amoebae within the cell mass rather than any kind of growth,
the repellant gas must in some way affect the rate of movement

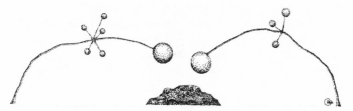

Fig. 21 Two rising slime mold fruiting bodies are attracted to a
small piece of activated charcoal on an agar surface. (Drawing by
R. Gillmor.)

of cells. If the gas is more concentrated on one side, the cells on that side will move faster, and the result will be that the tip of the cell mass moves away from the gas. Similarly, the absence of the gas on one side of a rising cell mass, caused by activated charcoal, makes those cells at the tip near it move more slowly, and therefore the tip points into the charcoal as though it were attracted to it. Finally, this explains the great puzzle of why small fruiting bodies rise at right angles from the surface, for in that way they equalize on all sides the concentration of the gas they give off, with the result that the fruiting body rises straight up into the air. The cell masses not only communicate with each other by assuring that they fruit in such a way that their spore heads are as far away as possible from one another, but they can orient themselves by keeping the gas they produce equally concentrated on all sides.

These experiments were done in the 1960s, and I have returned to the problem in the last few years. The remaining big question was: What is the gas? A student had collected the volatile substances given off by developing slime molds in a culture dish, and with the help of gas chromatography had shown that they gave off carbon dioxide, ammonia, ethylene, ethanol, ethane, and acetaldehyde in measurable amounts. Certainly these were the obvious first candidates, but how to test them?

One day I decided to try something new to test these gases. I took a small Petri dish which could be well sealed and drilled a minute hole (less than half a millimeter in diameter) in the middle of the top lid. Alongside this hole I placed a small agar block with the primordium of a fruiting body on it, oriented in such a way that it would fruit pointing out in a horizontal direction, parallel to the inner surface of the lid (fig. 22). The dish was then placed in a large closed container and the gases were introduced into the chamber. After a few hours the orientation of the fruiting body was observed. If moist air, or carbon dioxide, ethanol, ethane or ethylene in relatively high concentrations, was introduced, the fruiting body aimed directly for the hole above it. In a few cases it even went through the hole and up into the air above, although mostly it hit the rim of the hole. Presumably this is the case because none of the outside atmospheres repelled; rather, the repellant accumu-

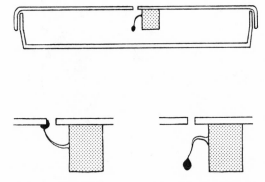

Fig. 22 The top diagram represents a section through a small plastic Petri dish (60 × 15 mm) showing the position of the agar block with the slime mold near the hole in the cover. These small Petri dishes are then placed in a large chamber containing various test gases. The lower diagrams are magnified views of the hole region. On the *right*, the fruiting body is repelled by ammonia; on the *left* is shown the response to all the other gases tested. (From Bonner et al., *Nature* 323 [1986]:630.)

lated inside the small Petri dish, given off by the slime mold itself, and for this reason the fruiting body tried to escape through the hole. However, if very low concentrations of ammonia (0.0036 percent compared to the 1 to 15 percent used for the other substances) were introduced into the outside atmosphere, the fruiting body would point directly away from the hole. The evidence indicated strongly that ammonia was the repellant, which was confirmed by other experiments in two other laboratories. It was an exciting moment when I first saw this—a moment of slime mold eureka!

The only fiasco occurred when we tried to test acetaldehyde. I had not realized that it had such a nasty, penetrating smell, and for a whole day the entire floor of the building had a nauseating reek to it. I felt very lucky that the solution to the problem did not leave us at the mercy of such an unpleasant substance.

If ammonia were the repellant, we had to be able to show that it speeded up the movement of the cells. This turned out to be relatively easy to do, and we were able to show the con-

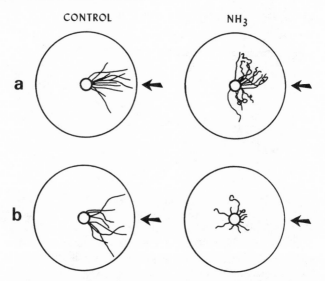

Fig. 23 The effect of ammonia on the orientation of slime mold slugs to light. Two experiments (a and b) show slime tracks that have migrated out from a central plug of agar in a Petri dish. The arrows show the direction of the light. Note that in an atmosphere of ammonia the slugs become disoriented and fail to move towards the light. This is only partially true in (a) where the ammonia is less concentrated than in (b). (From Bonner et al., *Proc. Natl. Acad. Sci. USA* 85 [1988]:3885.)

centrations of ammonia that speeded up single cells, aggregation streams, and migrating slugs, within the range that ammonia is known to be given off by the slime molds.

There was also the interesting possibility that orientation towards light, which is very striking in slime molds, especially during the migration stage, is also controlled by ammonia. The first experiment that supported such an idea was performed by a bright high school senior who was interested in doing some research. For a control she put a plug of agar containing incipient slime mold slugs in the center of a clear agar culture dish and shone light from one side. All the slugs oriented towards the light. If she repeated the same experiment with a sufficient concentration of ammonia in the immediate atmosphere, the slugs were totally disoriented (fig. 23).

Clearly the high concentration of ammonia gas wiped out the ability of the slugs to respond to light.

There is an interesting feature concerning the orientation to light that was first discovered many years ago by a German biologist. His experiments have been repeated for slime molds. The point is that slugs are translucent, and as the light hits their cylindrical shape, the slug acts as a magnifying lens and concentrates the light on the back side of the slug. This means that the light is more concentrated in the slug on the side that is farthest away from the source of light (fig. 24). A neat control for this is to put the slug in mineral oil, and then the slug goes away from light instead of towards it. This is because the refractive index of oil is such that the slug forms a diverging lens instead of a converging one, and therefore there is more light on the side of the slug nearest the light source. Both experiments indicate that the cells that have the greatest amount of illumination move the fastest. It was now a question of seeing if slugs migrating in the light produced more ammonia than the same slugs migrating in the dark, and the answer was unequivocal: light stimulates the production of ammonia, and we have seen that ammonia stimulates the speed of movement of the cells. Parallel experiments with slugs oriented in heat gradients, to which they are very sensitive, have given a similar result albeit with a few interesting complications.

From these simple experiments it is possible to see how a particular substance, which is the product of metabolism (largely the breakdown of proteins and amino acids), can become an important signaling molecule in slime molds. Ammonia is a small molecule that diffuses extremely rapidly in air and can penetrate cells with the greatest of ease. Furthermore, when it does get into cells it turns them more alkaline. We now want to know more about the details of how cells are speeded up by ammonia, and especially how such very small differences in the concentration of ammonia can make such large differences in locomotion. These are biochemical questions which are already being considered by others.

I have taken so much time and space to give many details of this primitive signaling between multicellular organisms for a number of reasons. First, it shows that even the lowest life

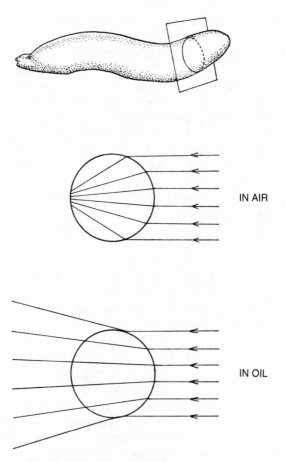

Fig. 24 Here we can see how light is focused on the far side of a slug in air, because the slug acts as a converging or magnifying lens. But if the slug is in mineral oil, the refractive index of the oil is such that the slug becomes a diverging lens. As a result, the slugs are attracted to light in air and repelled in oil because the side of the slug with the greatest illumination moves most rapidly.

forms have a very complex and sophisticated signaling system. Second, it lays out some of the basic principles of how signaling systems work. Finally, since the attack on this problem has spanned my scientific life so far, I look upon it with a special attachment.

• • • • •

Let me now take a great leap to the behavior of animals. The basis of that behavior also involves signaling, but animals differ from slime molds in that they have nervous systems. Some nerve cells are specially constructed to receive signals while others are specially constructed to activate cells which have some particular activity, such as muscle cells that are responsible for the movement of the animal. Between these two end points are intermediate or connecting nerve cells of varying degrees of complexity, the ultimate being a large collection to form a brain. These massive groups of connecting neurons not only modulate the responses to signals, but can initiate the signals on their own. They are message processors, and in their ultimate human form they are amazingly efficient thinking machines.

As we have seen, signaling is a means of communication between individuals and between individuals and their environment. For communication to occur between individuals, there must be a way of sending a signal and a way of receiving or registering that signal by another individual. It is certainly true that behavior is not just signaling, but every behavioral act involves communication of some form. Here I want first to outline briefly the various ways in which signals are used, that is, what the signals do for the interrelations of the individual animals. I will follow this with an examination of each kind of signaling, that is, visual, auditory, chemical, and so forth. Finally, I will give two examples in some detail of how signals can produce some remarkably complex behavior.

Signals are often used between species. For instance, alarms may be sounded by a prey species when a predator is spotted. Signals are also used to distinguish between individuals of one's own species, and those of a closely related species. If one knows one's own kind, for instance, this knowledge will be useful to avoid hybridization (with perhaps sterile results). In other words, signaling can serve as a means of isolating the individuals of a species from those of other species. Signals are also used extensively between the sexes. They are a means of selecting mates, marking and defending territories, and they are the basis of courtship itself.

Finally, signals are used between individuals of the same species over and above those related to sex. They are involved in all the complex relations between parent and offspring, for instance. As we will see in the next chapter, social animals depend on communication between individuals to produce a social group. Especially interesting are those signals which lead to the recognition of members of one's colony (nest odor in social insects) and, in some cases, to the recognition of individuals by the signals they give off. Let us now examine these signals and their analysis in more detail.

The modern era of our understanding of animal behavior came from the work of Konrad Lorenz, Niko Tinbergen, and a number of others. Their work blossomed and became widely accepted in the 1940s and 1950s, and this new discipline was dubbed "ethology." The most basic of the tenets of this new school centered around the significance of communication. They showed that organisms responded in fairly rigid ways to specific signals, and they called these responses "innate releasing mechanisms" to emphasize that they were rigid and inflexible responses. Signals were specifically tailored to the responses and were therefore called "releasers." One consequence was that it again became possible to think of organisms as responding instinctively to stimuli, something that had, since the last century, been considered an undesirable concept simply because instinct implied a fixed reaction, a fixed behavior. The idea that responses could be rigid somehow implied a determinism to behavior, and people found it difficult to divorce human behavior, with all its flexibility, from the behavior of other animals. It was felt that learning played such a key role in all behavior that there was a real danger in suggesting that there could be any kind of an automatic response that could be classified as a genetically inherited "instinct." The early ethologists made the notion respectable again, and today we have reached a compromise position. In many of the innate releasing mechanisms, learning may be added to the fixed, inherited response. The best way to understand the revolutionary ideas of these early ethologists is to give a few of their classical examples.

Sticklebacks are small, freshwater fish, in which the male marks off a territory and builds a nest. The females cruise about in groups, and as they pass a male, he will signal them by exposing his red belly as he swims up towards them. This bit of "flashing" often has the desired effect of luring one of the gravid females away, after which he leads her towards the nest by a series of characteristic zigzag movements (again signals and responses). Ultimately he will stimulate her to lay eggs, which he fertilizes. Subsequently he takes care of the young as the spent female wanders away to lead a carefree existence.

To show that this entire courtship consists of a series of stimuli and fixed responses, models of the males were made of painted wood. As a group of females passed by, the red wood model, imitating a real male, could be moved towards the females, one or more of whom would respond by following. In this way it was possible to see what aspect of the shape and color was important to stimulate a receptive female. The result was that the shape was of little consequence, but the red color was essential. Lorenz had an amusing anecdote about seeing a group of females respond in the total absence of a male. This puzzled him for some time until one day he bent down to the level of the fish tank and saw them do it just when a red truck passed by the window.

Another famous example comes from the work of Tinbergen on herring gulls. A newly hatched chick will instinctively peck at the red spot on the beak of a parent, with the result that the parent will instinctively regurgitate some food. Here are two fixed responses, one following the other. Tinbergen made models of beaks of various shapes with spots of different colors, and again the red color was most significant and elicited the greatest number of pecks from the newborn chicks. As the chick grows, it becomes more sophisticated, is able to learn more about the appearance of its parent, and is less likely to be fooled by a cardboard model.

This kind of learning is well recognized, but the ethologists found another unsuspected kind that they aptly called "imprinting." It is seen in young animals that are given one lesson, and if it is given at the right time they remember it forever, or at least through their childhood. The best-known example is an experiment with young ducklings. A few hours after they hatch, if they see a moving object of some size they will follow it. Normally that object will be their mother, but even if it is the boots or the trousers of a person, they will thereafter follow those human legs whenever they are near. One summer we bought two mallard hatchlings for our children, and they followed blue jeans everywhere! When the birds became larger they would follow us down to the river to swim (and they would not enter the water until one of us had gone in). Even when they could fly, if they saw us going to the river they would swoop down from the sky and laboriously waddle after us the rest of the way to the water.

. . . . .

Now that I have stressed the ubiquitous nature of signals and their responses and shown how they result in a variety of behaviors, we can proceed to examine the signals themselves and see how they elicit their responses. It is easy to classify signals, because they involve different kinds of energy of a physical nature. Visual signals involve transmission of light; auditory signals, the transmission of sound waves; chemical signals, the diffusion of molecules; and electrical signals, the movement of electrons. Each has fascinating examples which I will illustrate to show the richness of the signal-response system among animals.

It is obvious that any animal with reasonable eyesight can distinguish between other species and their own. Clearly, prey animals recognize the appearance of predators, and vice versa.

Much more interesting for our purposes here are those cases where two species closely resemble each other and the process of distinguishing between the two depends on subtle cues. An excellent example of this comes from some interesting work by Neil Smith on two species of Arctic gulls. They differ in the color of the eye ring: Glaucous gulls have a conspicuous yellow ring around the eye, while Thayer's gulls have a purple one. They breed in the same area, yet they seem to be quite able to avoid hybridizing with the other species. Smith asked whether or not the gulls differentiated the other species from their own by means of the eye ring which provided a modest visual cue, or was there some other method of discrimination.

He performed a clever experiment which involves two steps. In the first he brought together a female Glaucous gull with a yellow eye ring and a male Thayer gull in which the normal purple eye ring was painted yellow. The two birds took to one another and became a closely knit pair, but the male refused to mate with the female. Next Smith took the female and painted a purple band over her normal yellow one, and immediately they mated successfully and produced eggs. Obviously what was important was that they should see the eye ring color of their own species. Painting each member of the pair the eye ring color of the mate species gave both the illusion that they were courting their own species.

In another experiment Smith switched the eggs between a pair of Glaucous gulls and pair of Thayer gulls. After the young grew up they showed interest in mating only with their foster parents' species—not that of their genetic parents. In other words, the ability to identify with an eye ring color is learned—it is not an inherited trait.

There are innumerable examples of visual cues in courtship and mating. I have already described one in the case of the stickleback, but perhaps the most dramatic one of all is that of fireflies. Many of us have had the experience of taking a stroll at twilight on an early summer evening and seeing the fireflies (which are really small beetles) flash on and off. As a child I imagined them to be many Tinkerbells as they flashed like magic in the evening mist. What I did not realize then was that the flash is a signal for bringing the male and female together for mating. The male flashes and after a fixed time interval the female flashes back one or more times. She is often low on the ground or in a bush and her responding flash guides him to her. There are many species of fireflies all over the world—and often more than one species in one place—and each species has its own code of light flashes. For instance, there is a species in which the male gives three flashes in succession and then the female gives a responding flash, whose time interval is species-specific. In some species the male gives a long extended flash which elicits the female response flash after a suitable time period.

A most interesting series of discoveries of firefly behavior has been made by James Lloyd. First he showed that the females of some species respond not only to their own males, but to the males of other species as well, sending them her inviting flash. She has broken the code of the other species for a sinister purpose. The foreign males will swoop down on her, no doubt eagerly expecting sexual bliss, only to be devoured by the large predatory female whom Lloyd appropriately calls a *femme fatale*. But in more recent studies Lloyd has suggested that the males of some species send off false signals to test the female, and in this way they can discriminate between predatory females and ones of their own species. Even without this hypothesis it is clear that the fireflies' visual cues play a key role in finding mates (and sometimes finding a delicious meal). Some tropical fireflies, as a group, will densely populate an area and flash in synchrony. This is apparently a way of sending the message across larger distances, for the display is quite brilliant.

One of the commonest means of communication between animals is through the use of sound signals. This is something we humans are especially good at, for we have an extraordinarily rich language in which we can communicate all sorts of emotions, alarms, encouragements for others to come to us, and complicated thoughts and abstract ideas. In a comparatively simple and straightforward way, insects can also use sound signals.

Consider the lowly cricket calling its mate. A male will rub its wings together to make a cheerful sound, and when we hear it in our houses, we think it is a sign of good luck. Bringing good fortune to others is not exactly what the stridulating male cricket has in mind—he is seeking his own good fortune in the way of a mate. If a female is in the vicinity and she approves of his song, she will make as direct a trajectory towards him as she can, and they will mate. It is known that the male can produce only one song, and this specific song is the only one that stimulates the female. The song of a male of another species excites no interest: her response mechanism is specifically attuned to the song of the males of her own kind. The entire signal and response to the signal is firmly encoded in the genes; it is what is known as a fixed action pattern.

Birds provide particularly interesting examples of sound signals because they run the gamut from genetically fixed systems like the cricket's to ones in which the song is largely learned. Moreover, there is a good explanation for this variation.

Let us begin with fixed, inherited songs, such as that of the European cuckoo or the American cowbird. Both of these species are parasitic birds, that is, they perform the reprehensible trick of laying their eggs in the nests of other species. The parasitized parents do not seem to know the difference once the chicks have hatched. In fact, the cuckoo's chick will have such a wide and inviting gape when begging for food that the parent will cram the precious food it has collected down the cuckoo's craw in preference to that of its own less aggressive offspring. (The host parents do this because a big gape of a chick is an intense "releaser" to the adult birds and the parasitic chick automatically gets the rewards.)

The young cowbird or cuckoo will hear the song of its foster parents, but possibly never that of its own species. In the fall the young birds will migrate south, and when they return ready to mate they have never heard the call of their own species; they cannot have learned their own song. It is fortunate therefore that the males have inherited the ability to give their simple call, and the females have inherited the ability to respond to it by going to the male and assuming a position of readiness for copulation in an innate, automatic fashion.

At the other end of the scale are those male birds that learn their song from listening to their father. This is actually not quite true, for there are really three elements, as Mark Konishi showed some years ago. The bird (in his experiment, a white-crowned sparrow) does inherit an exceedingly crude basic song, which he greatly improves upon as he develops simply by hearing its his own song. Deafened birds can produce only the feeble inherited beginning of a song, and the crude song of isolated birds is vastly inferior to that of a normal male. With practice and by listening to others, a bird can turn its song into a work of art.

We know this to be the case with canaries. A young bird can be taught to sing like a master if it hears a master sing repeatedly over a period of time. (It is not always a success: we unfortunately had a canary that learned to imitate our telephone.) It is surprising how many birds can be taught to imitate a song. In the seventeenth and eighteenth centuries, caged birds in Europe were taught to learn simple songs played on a flute or a recorder. My research assistant takes care of wounded birds that need rehabilitation, and for some years we had a male starling in the lab whom she taught to whistle "Hail to the Chief." He was supposed to sing it when I came into the lab, but he was too smart for that bit of nonsense.

There are many birds that are quite extraordinary at imitating not only other birds, but also the human voice and all sorts of other sounds. Parrots, mynah birds, and mockingbirds are examples of experts in this category. Recently one of our sons told me that his neighbors had a small dog that drove them crazy with incessant yapping whenever someone came to the door. They decided to give the dog away, but a week before the dog went they acquired a parrot. No sooner had the dog gone when every time the doorbell rang the parrot gave the most perfect imitation of a frenzied dog.

There is the interesting question of why some birds are such good imitators. One suggestion is that mimicry is used in the formation of pair bonds in monogamous birds (where two birds can learn each other's signals), but I find that hard to imagine in cases where there is such extensive imitation, as in mockingbirds, for instance. There are a number of other reasonable hypotheses, and perhaps more than one of them is correct, depending upon the species of bird. I particularly like the idea that male song might be another example of sexual selection, and that elaborate song, imitated or otherwise, is the equivalent of gaudy plumage or elaborate bowers. This tactic substitutes sound signals for visual ones in the competition to become the most sought-after male. We do not know for a fact that the female picks the best singer, or that the star vocalist has the best and the richest territory, but it is a reasonable possibility. This would apply not only to the mimics, but to birds with beautiful songs such as the American thrushes, the European nightingale, and the winter wren that is found on both continents.

In competition between males, song is used as a way of marking the territory. The same is undoubtedly true for many mammals, as when dogs bark within their fence but remain quiet when they stray, or when howler monkeys use their great communal roaring to mark off the territory of their group in the jungle.

Another interesting use of sound signals occurs between parent and offspring, and a good example may be found in cliff-dwelling seabirds. Because of the great multitude of parents and chicks, they must know one another's cries so that the parents can feed their own chicks, and the chicks can identify their parents. It is now known that in murres, which live in dense colonies on ledges, the chick learns the calls of its parents when it is still inside the egg, so that as soon as it hatches it begs for food from its mother and father. The same phenomenon is found among mammals. I can remember once watching a score of ewes that left their small lambs and rushed to a trough the farmer had filled with some special mash. When it was all gone, the ewes and the lambs began to bleat and soon, through all the sounds, they found their own kin. Since everyone was making what seemed to me the same noise, it was quite startling to see how quickly they sorted themselves out.

It is important to note that in each of these examples of sound signaling (as was true in visual signaling) there is good reason to believe that the particular signaling system arose by natural selection. In each case it is possible to see a selective advantage to the particular behavior. Parasitic birds must inherit their song if they are to find mates; birds that learn song from their parents and other older birds can compete successfully for the attention of mates. The circumstances of each case suggest advantages that are often quite obvious and compelling.

Let me now turn to chemical signals. Insects provide many interesting examples because chemical signaling is one of their main means of communication. For instance, nest odor is a means of recognition among social insects, and in those cases where it was studied it appears that there are a number of chemical compounds involved, with their ratios no doubt differing from nest to nest. We could compare this to our ability to recognize different sauces made from the same ingredients, but some were mixed more skillfully than others. An insect, such as an ant, has an extraordinary number of glands which produce a variety of chemicals, often more than one in a particular gland. These glands are near the mouth, in the joint between the thorax and the abdomen, and in a cluster around the anus. They can produce alarm substances that announce danger to all the other ants in a colony; they have trail substances with which they mark the ground as they run along, perhaps leading other ants to a source of food. Edward O. Wilson has discovered a special substance used by slave-making ants. In raiding a colony of another species of ants to steal the unborn pupae for future slaves, they exude a particular substance that seems to send the adults of the victim nest into an absolute frenzy and they become quite hopeless at fighting the invaders. Wilson has called this a "propaganda substance."

In many ways one of the most remarkable phenomena among insects is the use of chemicals to bring the mates together. This is especially striking in moths of which the females emit a volatile substance and the male has huge antennae that are extraordinarily effective in responding to the chemical. Relatively few molecules (around two hundred) are enough to evoke a response, and once the male is alerted he will fly upwind until he reaches the female. The record for such an attraction is a case where the male was 6.8 miles from the female, yet he still found her in this fashion.

This great sensitivity to these sex pheromones is being used in pest control to attract the males of some species and then trap them when they arrive at a planted source of the attracting pheromone. Apparently those who do this kind of research are in constant danger of being pursued out-of-doors, because the attractant is so effective that it will stay in clothes even after they have been laundered in a washing machine. In one instance a technician from a laboratory in Seattle went to a football game, and to her dismay she became only too conspicuous as a cloud of moths hovered around her in the middle of the stadium seats.

There is a similar example among a mammal with which we are all familiar. Female dogs in heat can attract males from miles around by giving off a chemical signal during ovulation. This was brought home to me in a striking way at a time when I was teaching a large course in general biology as a struggling young assistant professor. We had a small spaniel bitch, and when she came into heat we felt we could not afford to put her in a kennel. But we soon discovered that not only was our dog the object of male dog interest, but so were we, for the attractant was inevitably transferred to us. Our small children were stalked in the garden by hounds twice their size. The climax came one day in the middle of my lecture class, when a huge Dalmatian and an even bigger Irish setter suddenly came streaking down the aisle of the lecture room, and both tried to mount me in front of the whole class. It was without doubt the most successful lecture I have ever given.

Because many mammals have such a keen sense of smell it is not surprising to find that scent is used for a number of purposes. A good example is the marking off of territory. Many deer have special glands, such as those near the eye or the musk gland on the abdomen, both of which are used specifically to mark off territory. Beavers have a gland near their anus which contains a scent called castoreum which they add to piles of their feces at the borders of their territory. This odoriferous pile, neatly patted into shape with their broad tail, is effective in warning encroaching beavers not to cross the border.

Canines in general, including domestic dogs, use urine as a

territorial marker. It is perhaps less obvious for our dogs, which live in a confusing human environment where they are forced to mix their rules with ours, but in the wild it is very clear. The Canadian author Farley Mowat, who wrote an entertaining book called *Never Cry Wolf*, noted from his cabin on the tundra that he was on the border between two wolf packs. Individuals from both packs would patrol the border and urinate at regular intervals along the edge. He decided to have a try himself, and after the consumption of a staggering amount of liquid and inhuman self-control he marked his own territory around his cabin at regular intervals with human urine, going into the territories of both packs. He was delighted to find his efforts were not in vain—the wolves respected his border.

For us, with our relatively insensitive sense of smell, it is difficult to understand how other animals can follow a track or recognize an individual by his or her odor. For many years we had a wonderful old basset hound named Phoebe. As I was walking to work one morning I saw Phoebe coming towards me on the same sidewalk, returning home from her stroll. I walked by quietly because I did not want her to follow me, but suddenly, after she had passed me by thirty yards, she gave a yelp, turned around in a frenzy of joy, and ran towards me; in her slow, absentminded way it took thirty yards to get my scent—not from me, but from my track. This is not unlike the experience of coming home after an absence, when your dog takes a sniff and gives you the most enthusiastic welcome.

This ability of dogs to recognize by scent has been explored in a very interesting way by geneticists. One of the earlier studies was by Hans Kalmus, who was able to persuade the London police force to lend him its most gifted tracking dog. He then gathered a group of about eight people, which included two identical twins, two fraternal twins, siblings, and totally unrelated individuals. He briefly put a handkerchief under the armpit of one of them, and then had all of them cross a field from one gate, crisscrossing one another's paths. The dog was then taken out of a place where he could not have observed any of this, given a chance to smell the handkerchief, and then ordered to track the person. He could easily discriminate between unrelated people, siblings, even fraternal twins. The only time he occasionally made an error was with the identical twins. The latter, by definition, are genetically identical, which means the dog could discriminate between genetically different individuals but faltered occasionally when they were identical. It is thought from some more recent work with mice that the scent identification is based on a few genes that are involved in the immunological system of mammals. What I find so hard to understand is not that there are a few proteins, recognized by the dog, that distinguish one individual from another, but that they can be detected through the soles of the shoes (or more likely wafted into the air), and are not confused when one track crisscrosses over another. I suppose if the moth can detect a female 6.8 miles away, this is no less a feat; but it is so foreign to anything we humans can do that no wonder I find it a bit of magic.

The idea that animals could communicate to one another by electricity was an idea that did not exist until the 1950s. It had been known for centuries that some fish, such as the South American electric eel, could generate huge shocks that would stun a prey or an enemy. These eels have a modified muscle which has evolved into a large battery, and just touching the fish will release instantaneously a huge store of voltage that is enough to stun a man if the eel is large. As a child I remember seeing them in zoos. The keeper would touch the eel with a prod that was connected to a row of light bulbs, all of which would light up. For some years I thought the sole purpose of electric fish was to light up bulbs in zoos.

At Cambridge University, H. Lissman began working on an essentially blind eel-like African fish called *Gymnarchus* that lived in impenetrably muddy water yet seemed to be able to navigate without difficulty. He showed that the fish gave off a weak, pulsatile electric current and suggested that it used this current to locate objects in the murk. The experiments were most convincing, but there remained a puzzle as to how the fish did this, for they not only had to generate the electric current, but be sensitive to it as well. This problem was solved a few years later when it was shown that, indeed, along the lateral line of a fish there are a group of nerve receptors which can specifically respond to small amounts of electricity. In the case of *Gymnarchus*, the fish is normally surrounded by a symmetrical electric field, and any distortion of that field by objects that are either good insulators or good conductors is recorded by the fish's receptors.

This use of electrical currents for location is hardly a case of communication. But once the system was discovered it was not long before it was shown that this electrical system is used in catching prey and is involved in communication between members of the same species. Some elegant experiments for the former show how a small shark finds flounders which are buried under the sand. If a shark comes near such a flounder it cannot see, it will still pounce and grab it. That the signal the shark receives is electrical can be shown in two ways. If the flounder is placed in an insulated chamber under the sand, the shark swims right by. If, on the other hand, an electrode giving off a current equivalent to that of the flounder is buried in the sand, the poor tricked shark will try hard to gobble it up.

Not all fish have electroreceptors or the ability to emit currents, but some of those that do can emit short pulses, and they are used in the same way as the light pulses of fireflies. Certain species of freshwater fish that live in small streams signal in this fashion, and their prospective mates respond with an appropriate reply. Not only is the signal given off at the right intensity, but the pulses are in the appropriate code. It is amazing to me that all these revolutionary studies have been made as recently as within the last thirty years. I wonder what other secrets of behavior of such a momentous nature animals are still hiding from us. Perhaps some species emit and respond to radio waves, or something equally unimaginable.

·   ·   ·   ·   ·

The first of my two examples of complex behavior has to do with honeybees. They are an ideal case because they show how much information can be extracted from simple signals, and that all this complexity can be handled by a lowly insect.

As in the case of electrical signals of fish, honeybee language is also a moderately recent discovery, made during the Second World War. The original work was done in Austria by Karl von Frisch (for which he received the Nobel Prize along with Lorenz and Tinbergen many years later); because of the war it was unknown in the United States, France, and Britain until some time later. We first learned about it from a well-known Danish physiologist who had nothing to do with the work but felt the Allies should know, so he summarized von Frisch's work in an article in a *Scientific American* of 1948. I remember it well, because many American biologists simply refused to believe it—the story was too fantastic to be true. He claimed that bees had a language, and one bee could tell others where to go look for nectar, even long distances away from the hive. To settle the matter, a respected zoologist from Cornell found some money to bring von Frisch over to this country for a lecture tour, largely to learn whether the story was true or pure fantasy, as it seemed to many. His lecture at Princeton made a vivid impression on me. Albert Einstein, then an old man, sat in the front row to hear von Frisch's story in a spell-binding lecture. After the lecture Einstein asked one of my colleagues if he could arrange to have von Frisch come to his house the next day. The arrangements were made, and my colleague, who was there during the visit, told me Einstein said to von Frisch that he saw a flaw in his experiments and suggested that additional ones were needed. Von Frisch replied that he had already conducted them, and they had also supported his theory. Einstein apparently was overjoyed, and they had a splendid time together.

What von Frisch had shown was that the dances given within the hive are those of returning scouts, and that the dance tells the other bees the location of the nectar. If a scout bee comes in and gives a round dance, a series of circles in which it first goes in one direction around the perimeter of a circle and then in the other, it means that the source of nectar is within 100 meters of the hive. Since the scouts also bring some odor cues from the flowers, the outgoing foraging bees soon find the nectar. (The scouts also mark the flowers with their own scent, to further ensure that the foragers find the rich source of food.)

If the nectar source is over 100 meters away, the scout bee is able to tell the other bees not only the distance with some exactitude, but also the direction. This remarkable feat, which was the source of much of the initial skepticism, was achieved by what von Frisch called the "waggle dance." In this dance the scout does a figure eight going one way and then the other, rather like a figure skater. It is not a true eight because the connecting zone between the two circles is a straight line, making it look more like two capital D's attached back to back. This straight line that borders on the two squashed circles is the all-important part of the dance. First the line points right towards the nectar and the scout bee always goes along the straight line in the same direction. As she does this she waggles her abdomen back and forth vigorously. The time she takes to go up the straight part corresponds to the distance the food source is from the hive; if she transverses the straight path in two seconds, the nectar can be found about half a kilometer away; if the transverse dance takes 6 seconds, the nectar will be 4 kilometers away, and all the times in between will correspond to the appropriate distances.

But this is not the whole story. The bees cannot really get an accurate direction from the short straight part of the dance alone. They also use the sun as a guide. To show how this is done, you can put a hive on its side so that it is horizontal (some primitive bees only dance this way). Then observe the direction of the dance and measure it as an angle from the sun. Suppose the dance takes place 30° to the right of the sun. When the foragers set off, they fly exactly 30° to the right, allowing them to keep a constant reference point, which would be necessary if they are to fly 3 or 4 kilometers away from the hive. If the food lies directly toward the sun, that's where the dance will point; if it lies directly away from the sun, again the dance will point accordingly.

However, the hives of honeybees do not lie on a horizontal plane but hang vertically. Furthermore, a scout may do her dance and give the crucial information about the location of honey to the other bees when it is dark. How is this possible? It is a remarkable fact that the scouts do this by translating a visual cue into a gravity cue. If the dance is performed straight up on the vertical hive, it means the source of food is directly towards the sun. The workers crowd around the dancing scout and essentially feel her directions in the dark. They then emerge from the hive and confidently fly directly towards the sun, stopping at the right distance because they also took note of the time required to waggle across the straight part of the dance. If the scout bee makes her straight line 30° to the right of the vertical, the foragers fly out and go 30° to the right of the sun.

One might wonder how the foragers can gauge the time it takes for the scout to transverse the straight path of the dance in the dark. This has been a subject that has been studied recently with interest. The waggles are quite rapid, but their frequency, that is, the number of waggles per second, is constant. To our ears the waggle makes a buzzing sound; it has been shown that this buzz gives off pressure waves which can be heard or sensed by the other bees if they are close to the scout. It has been possible to make a mechanical scout bee that gives off the right frequency of pressure waves, and the foragers will go out and follow the machine-made signals correctly.

In all this extraordinary story (and there is much more to it than I have given here) we see a whole set of different signals working together to produce this complex behavior. First there are the chemical cues connected with the round dance so that particular flowers rich in nectar can be identified. Then there are visual cues which are used to locate the sun and, of course, to see the pattern of the flowers once they are in the right area. Bees have quite an extraordinary ability to memorize landscapes. They will not only know the general geography of the area about the hive (which allows them to navigate on overcast days), but they can return to the exact location of their particular hive, even when it is one of many stacked together. Then there are tactile cues as the bees feel the scout with their antennae while she dances. There is also a kind of auditory cue in the form of a short-distance sound signal given off by the waggle dance. Finally, the bees can sense gravity, which along with touch and sound is due to some kind of pressing on strategically placed receptors that transmit the information to the brain of the bee.

Without doubt the most unexpected aspect of this tale is that gravity cues can be converted into visual cues and vice versa. It has often been said that the use of symbols is an achievement only human beings can claim. We now know that primates in general, and even lower vertebrates, may also use symbols for communication. It was hardly anticipated, however, that insects might also be highly skilled at using symbols, for that is exactly what the gravity cues amount to when one bee tells the others where to find food.

. . . . .

The question of how birds manage to migrate great distances and return to the same locality each year and the question of how racing pigeons manage to return home with such amazing accuracy and speed are matters that have intrigued humans for centuries. It is only in the last fifty years that we have begun to find some answers, but even today we do not have the complete story. There are undoubtedly more exciting discoveries to be made.

Migration requires cues for direction. In this case it is not signals between individuals, but signals from the outside world that the bird must read in order to orient properly. As we shall see, the moral to this story is that there are clearly numerous external cues the bird uses, and it is likely that there is no one answer to the mysteries of migration and homing, but many. Let me begin a description of each of a series of external cues that are important to varying degrees.

There are excellent examples which show that birds, like bees, are good at memorizing a landscape and finding home. My old friend Donald Griffin, who did so much to reveal echolocation in bats, performed an interesting experiment on this problem when he was a graduate student at Harvard University. He wanted to follow, in a small airplane, gulls he had removed from their nesting area. He did not know how to fly, and as a graduate student did not have the money for lessons or for renting a plane. Being a clever fellow he discovered that Harvard had, among its great resources, an unused fund that had been sitting fallow for many years and had accumulated just the amount of money he needed. It had been started by the great psychologist-philosopher William James to investigate the sixth sense, a proper field for investigation during a period when people genuinely believed that mediums could communicate with the dead. Griffin convinced the authorities that homing was beyond a doubt achieved by means of the sixth sense, and he was given the money. After he had learned to fly, he removed herring gulls from their nest, painted them bright red with a water-soluble dye to make them more easily visible, released them at various points, and followed them in his plane. Without going into the details of his results, I can say that he found that birds released near the coast returned along the coast to their nest—they knew the coast and where they were. The birds released inland would wander a bit, but once they found the coast they also returned quickly. This ability depends not only on keen eyesight but, as in bees, a wonderful memory for geography.

Undoubtedly the most important method of orientation is by means of the sun. Again, like bees, birds can orient with respect to a specific angle to the sun, but this raises a whole new subject.

Let me first state the problem in the simplest terms. A bird that is flying south in the fall flies a good part of the day. If it stayed at a fixed angle to the sun it would be constantly changing its direction as the sun moves across the sky. In fact, by evening it would have made a complete U-turn and would be flying north. For this reason it had always been assumed that the sun played no role in migration, but this view was turned upside down by the splendid experiments of Gustav Kramer, in Germany, who published his first results in 1950, the same year von Frisch first published his work on bees.

Kramer built fairly large circular cages and placed starlings inside them, with a perch in the middle of each cage. The birds would always flutter in the same north-south direction, as would be their expected route if they could see the sun. If

Kramer altered the apparent position of the sun with mirrors in a special room, the direction of the birds' flight was also altered, maintaining the same relation to the false position of the sun; on overcast days the birds seemed to lose their sense of direction entirely. If he used a light bulb in place of the sun in a room without windows, and if the light bulb was not moved around like the sun, the bird kept changing its direction—somehow it used the sun for a directional signal while also compensating for the normal change in position of the sun during the course of the day. The only possible conclusion was that the bird had an internal clock that automatically took into account the changing position of the sun.

The realization that animals and plants have a biological clock inside them has also come only in the last forty years. At first cycles were thought to be a property restricted to animals with a nervous system, and that they manifested themselves in human beings in the form of jet lag and menstrual periods. But such built-in clocks were soon recognized as a quite universal phenomenon, and plants and even unicellular organisms were seen to have daily rhythms that were governed by an internal clock. The chemical and physical bases of this clock remain obscure, although we do know some things about them and believe that the time will come when we will understand them completely. If the organism does have a nervous system, we know that the clock is controlled in a specific region of the brain; if that part of the brain is damaged, the animal loses its ability to cycle daily. (Recently it has been shown that with an implant in that region of the brain from another animal, the rhythm will be restored.) These clocks are used in a variety of ways so that an organism can fit its activities to the cycles of night and day or the cycles of the seasons, but here I want to show how the clock is used in the sun orientation of birds migrating in daylight.

A starling has a built-in system which automatically tells the bird where the sun should be at different times of the day. To prove this capability, Kramer put some starlings in a cage in which the daily light-dark cycle was shifted six hours, that is, dawn and dusk were six hours later than in the outside, real world. These birds were later put in a cage where they could see the sun, and instead of flying south, as they normally would in the fall, they flew 90° to the right, or due west. The reason is obvious: their clock tells them it is six hours later and therefore the sun should have shifted 90° in that amount of time (one quarter of its twenty-four-hour orbit). In other words, their angle of flight with respect to the sun is correct for them because their day-night cycle has been advanced six hours.

It should be noted here that bees do the same thing. They also have clocks which tell them how to fly at different times of day. This can be shown by capturing a bee that is going out to find nectar where the scout bee said it was. If the captive is kept in a dark box for 2 hours, it will continue in the correct direction despite the fact that the sun will have moved 30° in those 2 hours. If it had not used its clock it would have flown 30° off course.

Not all birds migrate in the daytime; many travel at night. One might imagine that they use the stars the same way starlings use the sun, but that is not necessarily so. In the 1970s Stephen Emlen, building on the previous work of others, did a fascinating series of experiments with indigo buntings, a beautiful bird that migrates at night. He was given permission to use the planetarium at Cornell University, and in the spring or fall he would capture migrating birds and put them in a cage inside the planetarium. The cage was cleverly designed, with an ink pad at the base surrounded by a cone of white blotting paper. The bird would stand on the ink pad and every time it made a leap or a flutter, the direction was automatically recorded by its footprints on the blotting paper. The night sky was rotated normally, and in the fall the bunting would leap south, and in the spring it would go north.

Emlen was able to demonstrate two radically new things about the way his birds oriented. In the first place, they did not pick one star or a group of stars and use them the way starlings use the sun; instead they used a fixed point—the North Star—because it is the only star in the northern sky that does not move, and all the other stars rotate around it during the course of the night. This means, among other things, that the indigo bunting does not use its internal clock. It does not need to because the North Star stays in the same place all night. Emlen's second discovery was that this trick of using the stationary North Star as a guide is learned; it is not directly inherited like the sun orientation of day-migrating birds. If nests were put in cages in the planetarium and the young chicks were allowed to see the sky rotating about the North Star every night, they would grow up to be capable of normal seasonal migration. If on the other hand they were reared in an ordinary room, they were incapable of flying in any consistent direction when placed under a normal night sky. This is another instance where one function (migration navigation) can be either learned or fixed in the genes.

There is one final point about bird migration that I find especially interesting. How does a bird, in the Northern Hemisphere, know to fly north in the spring and south in the fall? Apparently the hormones control the brain to make the appropriate response to the season. In the spring the sex hormones are beginning to surge in the body and they somehow tell the brain that north is the way to go. When those hormones are at their low ebb in the fall, the message is south. The question of how a major direction like this can be controlled by the genes, and how the hormones can reverse the signal, is a genuinely intriguing problem.

Many proposals in the past have suggested that birds are somehow sensitive to the magnetic forces which cover the surface of the globe, but they have always been dismissed (and with some justification) as crackpottery. It came, therefore, as a considerable shock when William Keeton, also at Cornell University, showed in the 1960s that homing pigeons could use magnetic cues for orientation. It is sad that Keeton died at an early stage in his career, and no one could have told the story of his discovery better than he.

As a boy he had a great interest in racing pigeons. As he became a scientist he was appalled to find that biologists who worked on homing were not trained in the proper care of pigeons. They treated them with about the same respect as white laboratory rats. When Keeton began his work, his first step was

to hire an experienced loft manager, as one is called in the pi-
geon-racing world, to keep the birds in perfect shape, allowing
them to race (i.e., experiment) only when the condition of
each bird was just right.

He found that pigeons on totally overcast days could home
quite well—something that had been known to pigeon racers
for many years, but not to biologists. He was also able to con-
duct the experiments on sunny days by fitting the pigeons with
opaque contact lenses, so that what they saw was equivalent
to a heavily overcast day. If the birds were fitted with a small
headset that obliterated any external magnetic cues because
they generated small ones that interfered, they became quite
lost on very cloudy days (or while wearing contact lenses).
Controls with the same headset but no magnetic generator
were not able to prevent accurate homing. The case was very
convincing. An added story Keeton used to tell was that pigeon racers in any part of the country know that there are certain places one should never start a race, as the birds would
become quite scattered and often got lost on cloudy days.
These sites had very peculiar magnetic anomalies due to the
local rock formations, something geologists had known for a
long time.

It is still not known how the bird senses the magnetic field.
Considerable excitement was generated when James Gould
and his associates showed that bees had small quantities of
magnetite in their bodies, and later it was shown that pigeons
have magnetite in their head. We still do not know how either
bees or birds read the magnetite compass, or whether in fact
that is the way they sense magnetic forces; the evidence is only
circumstantial, but the case is a strong one. Bees, incidentally,
are known to be influenced by external magnetic fields when
they orient their hives, but we have no idea what this or any
other role of magnetism might be in their life activities. Very
recently magnetite has also been discovered in human brains,
but there is no clue yet as to its function.

There are a number of other possible cues that might play a
role in migration. For one, sound could provide a cue. As a
bird flies in the air, the world around it is clear of extraneous

sounds, unlike flying in a noisy airplane. When it is cloudy it may hear the lapping of waves to tell it that it is near the coast. Or it may hear waterfalls, or frogs singing in the marshes. With the bird's exceptional ability to learn, these sound landmarks could play a role in its landscape orientation. Also, it is known that birds can hear sounds of very low frequency, below, for instance, our hearing capabilities. These ultrasounds have the advantage of traveling great distances, which means that sound cues from far away can be used.

It is also known that birds may be very sensitive to pressure. Pigeons blinded by contact lenses seem to know when they are approaching the ground, for their delicate pressure sensitivity means they have a built-in altimeter. There is an interesting story about the migration of small warblers from the New England coast to the Caribbean. In the fall they remain on the coast until a high pressure period of weather arrives with brisk northwest winds. Then they depart in masses, allowing the wind to help them over the great expanse of ocean. It could be that their cue to depart is the pressure rise, or it could be the favorable wind that carries them along. The uncertainty of how this happens only strengthens the idea again that there are many other forces, other cues, which are used in bird migration that we have not yet even imagined.

. . . . .

I have treated behavior from primitive multicellular organisms to complex animals as though it were an edifice built on a signaling system. The signals may come from one organism to another, but they also can come from the outside world to provide a basis, among other things, for orientation in the environment, as we have seen in this discussion of slime molds and bird migration. It is also clear that even though the signals and the responses to those signals may be elegantly simple, they can accumulate into a larger set of interweaving signals and responses that will produce a complex behavior, of which I have given two examples.

Remember that every one of these remarkable adaptations arose by natural selection, so that the adult stage of the life cycle could maintain itself and, more importantly, that it could do so in order to reproduce successfully. We have, however, not seen the end of the surprising things that have been produced as a result of the evolution of behavior. One of these is the evolution of animal societies.

Chapter 7

BECOMING SOCIAL

ONE OF the interesting things about life cycles is that they can combine in a way to form a social group which itself can cycle. Some have called such societies "superorganisms" to show the cohesiveness of the individual life cycles that come together in groups.

The study of social animals is called sociobiology, though it is difficult to define a social animal. We mean by "social" that individual animals interact behaviorally with one another, but all animals, especially vertebrates, do this to some degree. Consider, for instance, a grizzly bear or a tiger: it spends most of its life as a loner, so how could it possibly be considered social? But they are genuinely social, if only for short periods, for instance, when the males and females mate. An even more obvious example is in the relations and complex behavioral interactions between the mother and her cubs, which may extend for considerable periods of time.

We can easily extricate ourselves from the problem of deciding which animals are truly social by simply saying that all sexual animals are social, but that they differ enormously in degree. At one end of the spectrum we have the elaborate social activities of our own species and those of social insects, and at the other end we have bears and tigers, not to mention a multitude of relatively solitary animals from all other groups of animals.

But why are animals social? In particular, why do some have such very complex social organizations? I first became interested in these questions from my work on slime molds because they involve the interaction and coming together of many amoebae to form a multicellular organism. While writing a popular article about them many years ago for *Scientific American*, it occurred to me they could be called "social amoebae" and I used that as my title. This turned out to be a happy choice, for I can remember being asked a few years later to talk to some visiting Russian biologists on slime molds. They showed a total lack of interest in what I was saying, so I shifted gears and explained that they were social amoebae. Immediately their faces lit up, delighted to hear that socialism had spread so far down the ranks.

Today we think of all the degrees of social existence among animals as the consequence of natural selection. In other words, for some species there seems to be an advantage in forming a social group—an advantage, that is, in the Darwinian sense where it leads to greater individual reproductive success. I will now discuss a number of such possible advantages, but because one of them has produced so much controversy, a bit of the recent history of sociobiology is pertinent and illuminating.

An interest in social animals and their habits goes far back in the annals of natural history. Perhaps the most abundant literature centers around descriptions of social insects: honeybees, wasps, ants, and termites. Some of them form enormous groups, involving millions of individuals with an often complex division of labor.

The revolution in our looking for causes of this kind of animal society came in the 1960s when W. D. Hamilton suggested in a celebrated paper that there could be genetic reasons why a social existence might be advantageous in natural selection. Hamilton argued that close relatives shared many identical genes, and therefore acts of altruism were, from the point of view of the natural selection of genes, not suicidal but beneficial for the perpetuation of those genes. Hamilton showed that because of a very special property of Hymenoptera (i.e., bees, wasps, and ants), namely that the males have only one set of chromosomes (haploid) while females have the normal two sets (diploid), the female workers (all the workers are genetically female) are more closely related to one another, and less related to their offspring, than are siblings and offspring of most other organisms, such as ourselves, where both sexes have the double (diploid) number of chromosomes in their nuclei. We have the same probability of sharing the same genes with our siblings as with our children. The different situation in Hymenoptera is only true if all the workers have the same father, for if the males have only one set of genes, all the female offspring have the identical paternal genes; this is what makes the sisters so closely related to one another, having a probability of sharing 75 percent of their genes. For normal diploid organisms the chances of sharing a gene with a sibling is only 50 percent. Hamilton suggested that this close relatedness of the workers in one colony could account for their remarkable cooperation with one another, and their apparent self-sacrifice. He also suggested that since only the Hymenoptera had this peculiar difference between males and females, one ought to find more colonial forms among them than in other social insects such as termites, where both sexes are diploid. It has long been known that ants, bees, and wasps have evolved or invented a social organization at least eleven separate times, while termites invented it only once, an argument that supports Hamilton's kin selection hypothesis.

It is an interesting bit of history that even though Hamilton published his paper in 1964, its full implications were not appreciated at large until E. O. Wilson published his encyclopedic blockbuster of a book called *Sociobiology* in 1975. It is a book full of riches on social animals at all levels, written with intelligence and clarity. Among many other things, it explained Hamilton's proposition so effectively that it became widely known and understood; furthermore, Wilson explored not only how it applied to social insects, but to other animals as well. I was and remain an ardent admirer of the book, and reviewed it very favorably in *Scientific American*. But along with Wilson, I was politically naive and totally failed to anticipate the massive attack by a group of Marxists that was launched immediately after its publication. The Science for the People group in Boston wrote a vitriolic polemic against the book on the basis of some remarks in the short first and last chapters, which state that animal societies might tell us something of the nature of human societies. The group decided that Wilson was proposing that all behavior was determined by genes, something that would be anathema to Marxists, and in any extreme form most unsettling to any normal person. It was the virulence of the attack that took us all by surprise; it was as though Wilson had unwittingly provided the perfect excuse for an intellectual lynching. The newspapers had a feast: *Time* magazine ran a headline which said "Genes über alles," and at a national meeting someone poured a bucket of water over Wilson while he was on the platform before a large audience. In all this, the great significance of the idea that there could be genetic reasons for becoming social among animals, and all the fascinating detail of how animal societies work, became utterly lost. Along with most people I am not in favor of the idea that there is no free will among human beings, and that all our actions are genetically determined; yet I do want to discuss genetic relatedness and kin selection and how it might have played a role in the natural selection and evolution of animal societies. I will do this by avoiding the tender subject of *Homo sapiens* and confine my remarks entirely to nonhuman animals. (Later in this chapter I will have something to say about human sociobiology in the hope that I can do so without being politically assailed. I am told by my Marxist friends that this is impossible, and we shall soon see to what degree I succeed.)

The study of social animals, greatly stimulated by the works of Hamilton and Wilson, continues to be an exceedingly active and important field today. To begin with, there have been significant advances in our understanding of how selection acts and is correlated with kin relatedness in social insects. There is a great variation among insects in the size and complexity of their social structure, and these different levels have each provided insights into how the workers and the queen interact and cooperate. In general, the Hamiltonian principle of kin selection has not only held up, but has become greatly enriched in detail as well. However, it is also true that it has become increasingly evident from many of these studies that genetic relatedness cannot be the only factor for becoming social, and in many instances the primary cause might be something quite different. More on this presently, but first I want to discuss diploid social organisms, which is the rule for most organisms; the haploid males of Hymenoptera are the exception that quite literally proved the rule.

Let me give two examples where it has been argued that kin selection plays a role in the evolution of social cooperation in diploid vertebrates. One is a much-studied case of the Florida scrub jay, which has a behavior that is characteristic of numerous social birds. When a mating pair of birds builds a nest and raises young, two or three other mature birds often act as helpers. These birds are the offspring from the previous year, and even though they are capable of reproduction, they stay with their parents and help defend the nest and feed their newly arrived siblings. The nests that had helpers were more successful and reared more offspring in one season than nests without helpers. So here we see an example where kin benefit one another, and do so by sacrifice, by being nonreproducing helpers, yet many of their genes are shared by the siblings they protect and feed. However, we must remember that although they increase the survival success of their siblings, the helpers are not reproducing themselves, which would be the most obvious way of perpetuating their genes; the genetic gain derived from helping is counterbalanced by an appreciable loss. This realization has led to work which shows that the year of helping by the young birds is not all altruism, but that this is a year of apprenticeship on how to be parents. This extra time of learning equips them, when they later go off on their own, to get better territories and to be more effective parents, rearing more young than inexperienced birds. This tale clearly shows that there are many facets to the question of how kin selection could favor cooperation, but the basic principle remains unshaken.

My second example comes from African lions. A pride of lions is mainly an aggregation of females and their cubs, plus one or two males. The males, however, lead a precarious existence for they have to defend their position in the pride from marauding males that challenge them. The fights that result for supremacy over the pride can be deadly, and if the victorious male (or males) are from the outside they will, in a relatively short time, kill all the cubs. These observations were explained with the argument that by cub killing they stop the flow of the genes of their rivals into subsequent generations and proceed to sire their own cubs. In human terms (in which I should not indulge) they act like pathologically selfish stepfathers. It is also particularly significant that if there are two males instead of one, they are usually brothers; in other words they themselves share many genes and therefore are tolerant of each other's offspring. This example of infanticide by newly arrived males is found among numerous other social mammals, particularly among primates. As before, such behavior fits in with the notion that one of the driving forces for a social existence in some animals has been kin selection, but I again caution that it is only one of the factors driving selection towards increased social interactions.

．　．　．　．　．

There are a number of other advantages to becoming social which I would now like to consider carefully. These include advantages gained in the gathering of food and in the protection of individuals against predators and from the harshness of the environment. Finally, in some instances, bringing the sexes together for reproduction has been facilitated by making courtship and mating a social event, sometimes a quite elaborate one.

It is increasingly obvious, contrary to one's natural intuition, that in their own way insects seem to be able to do everything that mammals and other vertebrates can manage. Therefore, it is not surprising to find examples of social insects gathering food in a way that would be impossible without involving great numbers of their members. Such is the case for army ants found in the tropics in the Americas, and driver ants in Africa. They form great marching columns a dozen or more ants in width, extending great distances as they file through the jungle floor. One of the characteristics of these ants is that they love to follow, but hate to lead. As a result, the front end of a column spreads out in a fan, because as soon as an ant is in front, she slows down, waiting for the hoard to pass her and allow her composure to be restored. The result is a widening fan of reluctant leaders that is well suited to finding prey. Should some unwary worm or fat insect grub be discovered, even if it is many times larger than an individual ant it will be literally torn apart by the great multitude. There are many wild stories of how army ants might find a tethered donkey, leaving nothing more than a pile of bones by morning. However thrilling a tale this is, it is not true, even though it does illustrate in a most exaggerated fashion the point I am making. It is true, however, that the march of driver or army ants into a house is considered a desirable event, because they will either devour or chase away every form of vermin, especially cockroaches and mice. It is said that in places where such visits are likely, it is wise to put the four legs of one's bed in pans of water in case the throng arrives when one is sleeping, for their mass biting can be extremely disagreeable.

A number of social mammals hunt in packs, and in that way can fell much larger animals than if they hunted alone. A good example is wolves attacking moose, of whom they kill the smaller, weaker animals. In Africa, it is the hyenas and wild dogs that work in packs. As in army ants, the packs not only may have methods of outflanking the prey, but once they have it cornered they can overwhelm it by the sheer weight of their numbers. Wolf packs are clever at dividing into two groups, one group driving the deer or sheep into the fangs of the other. Hyenas seem to just bear down on an unwitting victim by sheer speed and force and then dispose of the carcass, bones and all (with their powerful bone-cracking teeth), in a matter of fifteen or so minutes. Presumably this unseemly haste is to get the food inside them before the lions appear.

Clearly these social groups have acquired skills that would be denied them as solitary animals: they can eat animals larger than themselves. It is easy to imagine how the chance cooperation of a small family could evolve into larger and larger groups with ever increased efficiency in finding food. The nonsocial relatives of such species would concentrate on the smaller prey, while the social animals, by cooperative hunting, have found a new and unexploited source of food.

I can remember as a child seeing a dramatic drawing in an old German natural history book of the royal chamber of one of the more elaborate termite colonies in Africa. There was a gigantic queen—so distended that she looked like a great, pasty blimp, essentially incapable of movement. Around her were the small workers, scrubbing her sides and removing her eggs as they popped out with amazing frequency and regularity. The small king was walking by her side, looking rather lost and bewildered, and around this domestic scene was a perfect circle of ferocious-looking soldiers, with huge jaws, all facing away from her to intercept the enemy. Later I was greatly disillusioned to learn that the soldiers did not stand guard like this all the time, but only in the presence of danger. In this case it was the artist who poked a hole in the side of the chamber. Still, this remains a perfect example of protection by a group of animals, in this case performed by a special soldier caste with lethal jaws.

The very same phenomenon can be seen in musk ox and elephants. If musk ox are pursued by a pack of arctic wolves, their worst tactic would be to run away, which often has the serious consequence of losing a calf. If they are not panicked they will form a perfect circle, with their massive horns pointing outward in an impregnable shield. It is the principle used by Caesar in his Gallic wars, when he would place soldiers in a tight group with their shields facing the outside. Elephants do the same thing, although with somewhat less military precision. Elephant groups are made up only of females and their offspring. When alarmed they too form a circle of adults, all facing outward, with their trunks periodically pushing the restless young back into the center of the circle.

Such organized groups are effective in protecting the young—more effective than a solitary mother. Cooperative protection mechanisms are under strong positive selection pressure, and it is easy to see how they could have arisen by natural selection: the offspring successfully protected are the ones that carry the genes to the next generation.

Perhaps the greatest enemy of all animals is the elements. For the most part animals cope by means of remarkable physiological tricks to compensate for extremes in temperature and other adverse conditions. Here we are concerned with social methods of coping, that is, methods which involve a massive and concerted effort among numerous, cooperating individuals.

The best examples can be found among the social insects. Termites, for instance, have little power to resist desiccation and in a dry environment they will shrivel up and die. To protect themselves they build their nests, or termitaria, by sealing them off from the outside world, to keep the moisture inside. If they find some delicious wood far away from their nests, say over an inedible concrete floor or wall, they will build a little covered pathway or tunnel with wood pulp and saliva and extend it, often a great distance, to the new source of food, thereby staying permanently indoors in a suitably moist environment.

Honeybees are enormously skilled at controlling the temperature of their hive. During the active honey-gathering season they keep the hive temperature constant within one degree Celsius ($34.5°$ to $35.5°$ C). To test the bees' skill at temperature regulation, hives have been put in locations that are at the freezing point, that is, $0°$ C, and at the other extreme of $70°$ C (which is $158°$ F), and they can still keep the hive temperature within that one-degree limit. How can they manage this remarkable feat? If the outside temperature is on the cold side, they rapidly contract and relax their well-developed wing muscles and in this way release heat, which warms the hive. In hot weather they use a clever trick: they carry in and regurgitate droplets of water which they place near the entrance. Then they fan these droplets to encourage their evaporation, and the loss of heat due to evaporation cools the air in the hive. It is the same principle that allows the water to evaporate from the tongue of a panting dog. If many of the worker bees participate, the cooling effect can be considerable, providing perfect air conditioning on a hot day.

There are many other ways that animals control their environment by social activity. The beaver is a classic case. By cooperative building of a dam, beavers produce a shallow pond which will cover the entrance holes of their lodges and allow them to carry and bury food in the form of branches for winter storage under the ice. Their modification of the environment thus serves two functions: that of protection, and of the care and preservation of food.

Human beings are by far the most adept at modifying the environment to suit our needs and comforts. We range from using clothing that will suit weather conditions, to displaying our full cleverness in the inventions of central heating and air conditioning. We also build fortresses for protection and have many tricks (and keep inventing new ones) to modify our environment to suit our purposes.

. . . . .

Earlier I pointed out that the degrees of social existence run a great gamut—all the way from solitary bears to the highly social insects or mammals such as ourselves. I will now describe and compare a number of different social animals that can differ in various ways. Instead of going into great detail for each group of organisms, I will compare societies by examining two properties which are found in all: (1) how animal societies divide the labor and how this is related to the size of the society, and (2) how animals became effective in communication as they became social; signaling between individuals is a keystone to successful social organization.

As animal societies increase in size, there is a greater division of labor, which is another way of saying that they become more complex. In animals that have a minimal social existence, there is no striking division of labor, though there is some of a minor sort: a separation of the two sexes, and a separation of ages (between, say, the cubs and their mother). But these divisions are the direct consequence of the life cycle of all sexual organisms and impossible to circumvent. They are the biological givens that provide the initial basis for a social existence.

Looking at the other end of the social scale, imagine some species of ant that has more than a million individuals within one colony, one social group. Again there will be a reproductive female, but only one (the queen), and a male who participates in the nuptial flight and dies before the colony is formed. His sole legacy is his sperm, which the queen carefully preserves in a sac that leads into her oviduct. The queen lays increasingly vast numbers of eggs that form larvae which grow and pupate into adults. These are the workers; they are genetically female, but they are sterile. The remarkable thing about them is that in very large colonies not all the workers are the same size or the same shape and, furthermore, the tasks they perform in the colony are different. The very small workers specialize in attending the queen and rearing the larvae; they are the maids and the nurses. The middle-sized workers mainly forage for food. The largest workers are the Amazon soldiers, the guards.

There has been great interest over the years in how these castes are determined. We now know that workers of different size are genetically similar—their individual size is not inherited. Instead, these differences arise in two ways: one is by the amount of food they get as larvae, and the other is the production of "social hormones" (more properly called pheromones), which are chemicals passed among individuals. For instance, one caste may limit its numbers by producing an inhibitor that prevents other castes from developing into its own kind. This latter phenomenon is well illustrated in the termites, which develop in a manner slightly different from ants and go through a series of molts. A wonderful old experiment has made it clear that the large soldiers can actually inhibit other workers from developing into soldiers. This was shown in two ways. If all the soldiers were removed from a colony, new soldiers emerged in the molts that followed and reestablished their ratio to the other workers, identical to that of the original colony. These observations were followed up by grinding up the removed soldiers and making a paste of them, which was fed to the workers. In that colony no new soldiers appeared. Clearly there was a soldier pheromone that inhibited the workers from molting into soldiers. It is by means of such inhibitors that the proportions of the different castes in the colony are maintained.

It had always been thought that insects were the only group of organisms that possessed such highly complex and sophisticated societies and their extreme division of labor. By comparison, most mammal societies seem simple, largely because all the individuals in most social groups are physically similar; it was always assumed that there was no mammalian (or vertebrate) equivalent to the castes of social insects. However, this has turned out not to be true. In burrows under the ground in the more arid regions of Africa and the Middle East, there is a rodent called the naked mole rat which has a society that shows remarkable parallels to that of advanced social insects.

Without any doubt these mole rats hold the record for the ugliest little mammal. They are hairless, and their wrinkled skin varies from gray to pink-yellow. They have beady little eyes and two great, curved incisor teeth which stick out beyond their pursed lips. Fortunately they are rather small, as the name "rat" implies, so we are not likely to be frightened by one on a dark night.

Since they live underground they are difficult to study, but recently J.U.M. Jarvis of South Africa has been most enterprising in doing so. She brought an entire colony into the laboratory and was able, over several years, with the help of observation windows in the tunnels, to discover their extraordinary social system.

There were about forty individuals in Jarvis's colony. They could be separated into three main size classes, which she categorized as follows:

1. The small "frequent workers" (average weight 28 grams) build the nest, forage for food (roots and bulbs which can be attacked from underground), dig the tunnels, and transport the soil and food (which, with slavish restraint, they do not eat while transporting).

2. The "infrequent workers" are larger (average weight 35 grams) and perform the same chores as the frequent workers, but they are lazy and work only 25 percent of the time. Loafing is their main occupation.

3. The "nonworkers" are the large, royal members of the colony. They consist of a few large males and females (average weight 47 grams) all of which are capable of reproduction, but only one female is the true queen; by aggression and possibly by producing an inhibitory pheromone she sees to it that she is the only one that reproduces. All that these large individuals "do," besides the reproductive activities of a selected few, is to lie around, waiting for food to be brought to them.

There is something so outrageous about these beasts that one almost wants to organize the infrequent workers into a union and promote a revolution, but when I think how extraordinarily unattractive they are to look at, my enthusiasm for mole rat democratic reform wanes. However, when I succeed in looking at them as a biologist, I am uplifted. Here is a mammal that has a social system remarkably similar to that of termites. Clearly the whole system, with its castes, one actively reproductive female, and its division of labor among nonreproducing workers, has arisen independently in both groups. One then wants to know why this happened: what are the features of their societies that have been culled by natural selection? If we look for common denominators, we note that both live in underground mansions, closed off from the unfriendly, arid environment. This need might have been the reason for the initial group-living in both cases; they came together to build a collective protective shelter. But why then should both produce castes and have a division of labor? One can say that the mere size of some termite societies is so great that the division of labor is an inevitable consequence—efficiency in this form amounts to positive natural selection. It is also true that species of insects with smaller colonies have castes that are less conspicuous. However, the naked mole rat colonies are quite small, yet they still show rudimentary castes. Why castes are formed in this case is an interesting problem that requires more knowledge and understanding.

In animal societies of intermediate size and complexity, such as wolves and lions, there are no morphological differences between the individuals, yet they do have a division of labor. For instance, in wolves or African wild dogs, only one female and one male will normally be reproductively active, yet other individuals will serve as helpers in rearing the pups. The roles the individuals play in this society are behavioral, and with age and experience they can become the active breeders, themselves eventually to be replaced as they become old and feeble.

The relation that is established in such a social group is called a "dominance hierarchy." It is also called a "peck order" because it was first described in chickens, among which the dominant bird in a group pecks all the others and is not pecked by any other member. It was also shown that if a group of chickens who did not know each other are put together, there is initially a great mixup of who pecked whom, but with time they sort out to a linear sequence: A pecks B, B pecks C, C pecks D, and so forth. This principle of establishing a peck order is common to all social animals—it has even been established recently for the workers of ants. Wolves and wild dogs establish their modest division of labor in this way.

In primate societies the conspicuousness of the dominance hierarchy varies enormously. At one end of the spectrum are the African hamadryas baboons, whose smallest social groups consist of four to ten individuals. For repugnant male chauvinism they are champions: the top male will not only keep any other males in the troop from copulating with the females, but he will keep the females close to him in a manner that can only be described as savage. If one of his harem should stray too far away, he will chase after her, bite her fiercely in the neck, and make her scamper back to the fold. By contrast, howler monkeys from Central and South America have larger groups—often up to thirty or more individuals—and it is very difficult to detect which is the dominant male for they all seem to be living in good-natured harmony. If a female comes into heat, instead of fighting for her the males will calmly take turns. Relations among individuals are an ideal example of what it means to be "laid back." They are aggressive when they encounter a troop from a neighboring territory that encroaches on theirs, or even when potential predators appear. I can well remember my first encounter with them in Panama. As a very young neophyte I was walking through the forest, quite igno-rant of the fact that I was under a silent band of howlers. They suddenly began to roar all at once, producing violent palpita-tions that would undoubtedly have led to a heart attack had I not been so young. And before I recovered I realized that they were all, in a very effective group effort, excreting in every conceivable way right over me. I wished I had brought a large umbrella.

Dominance hierarchies are behavioral and they involve communication between individuals, that is, visual, chemical, or auditory signals. Clearly the howler monkey communicates with its enemies by producing a deafening sound, while indi-viduals communicate within their own social group by means of subtle grunts, various body and facial signals, and smell (to recognize individuals and females in estrus). Communication is absolutely essential for any kind of social group. It is not only the basis of dominance hierarchies, but every cooperative and aggressive act within a social group involves communica-tion. This is true of insects as well as social vertebrates. The small brains of bees, ants, wasps, and termites do not inhibit their extraordinary ability to manage large amounts of com-munication among individuals.

Another important example of social communication is the ability to recognize other individuals within the social group. Insects recognize nest mates by the nest odor that is carried by everyone in a particular colony. One presumes that insects come close to recognizing individuals too, because this must be true to some degree in ants that have a peck order within the workers of the colony.

Clearly birds and mammals are very much better at it than insects. Many birds, such as geese and swans, are monogamous and stay with one mate for life. The cues they use are both visual and auditory—few birds have any sense of smell. Many mammals, on the other hand, rely heavily on smell for recogni-tion. Other mammals, such as ourselves, rely far more heavily on sound—so easily illustrated by recognizing a familiar voice over the telephone—and especially on sight. We not only recognize faces with amazing accuracy and speed, but we can identify close friends even at a great distance just by their posture and the way they walk.

. . . . .

In recent years it has slowly dawned on many people that an idea of a few earlier pioneers (Alison Jolly, Nicholas Humphrey, and others) is indeed of central importance in the evolution of the societies of primates and human beings. The idea is very simple: through natural selection social groups have been enormously successful, a fact reflected in the large numbers of animals, both insects and vertebrates, that are social. In vertebrates especially, one of the best ways of being successful in reproduction is by having a superior intelligence; therefore, hand-in-hand with the evolutionary success of a social existence, there has been a selection for intelligence. We shall see how these two properties may have worked together to produce our own elaborate and complex society.

The first question to ask is, why are human beings so smart? Why has there been a selection for large and clever brains, a gift that fills us with wonderful self-satisfaction as we compare ourselves to all other animals. One of the standard arguments is that our ancestors found the use of tools especially rewarding, and their manipulation led not only to an increase of physical dexterity in the hands but also to an increase in the capacity of the brain to effectively use and govern tools of increasing complexity. This is indeed a reasonable hypothesis, yet it is hard to understand why these simple maneuvers with our hands made us quite as intelligent as we are; we can manage much more with our thoughts than the use of mere tools.

A far more interesting suggestion, made by Nicholas Humphrey and others, is that our intelligence has developed to manipulate individuals of our own species, for those that are most successful in this manipulation of their peers are the ones who are most successful in reproduction. For this reason the process has been aptly dubbed "Machiavellian intelligence."

The idea is a direct extension of the role of dominance in

producing breeding individuals in any social group of animals. Put on a human level, we fortunately are no longer a society where the greatest bullies are the dominant individuals; more likely dominance is established by the males and females with the greatest craft. They are the ones who can steer their less well endowed colleagues and friends to best serve their ends. At first it seemed clear to me that this could be a strong force in human society, but I had greater difficulty in appreciating how craft might play a role in primates and therefore in early human evolution.

The reason for this was sheer ignorance on my part of primate behavior, but I found all the answers I could possibly want in reading a remarkable book by Franz de Waal called, most appropriately, *Chimpanzee Politics*. For fifteen years de Waal and his students studied in great detail a large group of chimpanzees in a big compound in the Arnheim Zoo in the Netherlands. It was a revelation to me to learn how subtle and intricate are the bonds and the relations among individuals, and how they mentally (and occasionally physically) spar with one another. Let me give a series of examples that show how sophisticated the behavioral interactions are among social animals, especially chimpanzees.

What strikes one in the Arnheim Zoo, and indeed in the wild chimpanzees in the Gombe preserve in Africa, followed so diligently by Jane Goodall and her associates, is that the chimps can become very aggressive towards one another and sometimes end up having violent, physical fights. The interesting thing, carefully documented by de Waal (he has devoted a more recent book, *Peacemaking among Primates*, to this subject alone), is that after these fights, no matter how wild or severe they are, there must be a reconciliation. Once the adrenalin level of the individuals has subsided, they appear as though they cannot bear the thought of remaining at odds with their adversary, no matter how keen their rivalry might be. These fights are especially prevalent among the large males, and once the fight is over they will remain motionless, avoiding each other's eyes. After some time, one of them, usually the subordinate one, will suddenly extend his hand towards the other. This will be immediately interpreted as the moment to forgive, and they will hug each other and may even start to groom.

De Waal cites a wonderful example, which I will present in his own words:

> People who work with chimpanzees know from their own experience just how strong the need for reconciliation is. Probably no other animals species demonstrates this need so forcefully, and it takes some getting used to. Yvonne van Koekenberg has described her reaction to her first experience of this phenomenon.
>
> Yvonne had a young chimpanzee called Choco staying with her for a while. Choco was becoming more and more mischievous and it was time she was checked. One day when Choco had taken the phone off the hook for the nth time, Yvonne gave her a good scolding while at the same time gripping her arm unusually tightly. The scolding seemed to have the desired effect on Choco, so Yvonne sat down on the sofa and started to read a book. She had forgotten the whole incident when suddenly Choco leapt on to her lap, threw her arms around Yvonne's neck and gave her a typical chimpanzee kiss (with open mouth) smack on the lips. This was so completely different from Choco's usual behavior that it must have been connected with the scolding. Choco's embrace not only moved Yvonne, it also gave her a deep emotional shock. She realized that she had never expected such behavior from an animal; she had completely misjudged the intensity of Choco's feelings.

Another striking way that chimpanzees show the complexity of their interrelations with other members of their group is seen when one individual displays a certain behavior only in the presence of a particular individual. For instance, two of the males had a fight and one appeared to be slightly hurt and was limping, but he limped only when the other male was looking at him; as soon as he looked elsewhere the limp miraculously disappeared. Of course, this kind of behavior can be seen in other animals as well, which brings to mind an account of two Western psychologists who visited a student of Pavlov. She had to leave them for a bit and told them if they went out for a stroll, not to take her German shepherd near the staircase because he was terrified by it. Of course, no sooner had she gone that they had to test this phenomenon; the dog, however, trotted down the stairs without the slightest hesitation. Clearly the terror-of-the-stairs act was solely for the benefit of his mistress.

A far more important illustration of the role of Machiavellian intelligence is de Waal's demonstration of alliances within the social group and how they lead to dominance and power. There were three large males in his group of chimps, and success for any one of them depended on the support of at least one of the others. This resulted in endless maneuverings on the part of the male most anxious to have at least one supporting friend. He also had to be ever vigilant that the new friend did not turn the tables and take over by making an alliance with the third male. The palace *coups d'état* of small (and sometimes large) nations seem to be no different in their complexity and subtlety. Furthermore, a particularly important group in establishing an alpha male may be the females. They sometimes will take a fancy to a particular male and a dislike of another, and that settles the matter: their favorite wins, even if he does seem less strongly built and masterful to us.

People have long been puzzled about the reasons for dolphins' large brains. Extensive experiments have been performed to see if their whistles and grunts are some sort of elaborate language that we have been unable to fathom. A possible answer comes from some very recent work on bottlenose dolphins off the coast of Australia. The males cruise around in small groups, which are really alliances. One group may solicit the help of another, and together they might gang up on a third group that is escorting a receptive female. In other words, dolphin politics may be even more elaborate than chimpanzee politics, and this could be a possible reason for their large brains. This example is a bit of independent evidence that there has been a selection for Machiavellian intelligence.

To return to chimpanzees, let me give a final example of the cleverness of their interpersonal relationship from the work of Emil Menzel, who studied a group of chimpanzees in an open compound in Louisiana. He did an experiment in which he put all the animals except one in a cage, and he allowed that one to see where he buried some bananas before locking him up too. When the group was released, all of them followed the knowledgeable chimp who seemed to know where he was going. The next time the experiment was repeated, the same scout chimp, with his secret knowledge, rushed off in the *wrong* direction, and of course no one found any food. Only when he was fully satisfied that everyone was preoccupied did he sneak off and have a private feast on the hidden bananas. There are many examples of deception among animals, but this one is especially to the point, for it is a clear case of one individual manipulating the other members of its group.

. . . . .

My argument has been that there are many factors that have led to a social existence among animals: mutual protection, more effective food harvesting (often with the ability to obtain food that would be impossible for single individuals to get), conquest of the vagaries of the environment by the creation of a stable colony environment, and social cooperation for the sake of preserving genes held in common with relatives. No matter what the forces of natural selection were that led to social groups, be they one of the above or some combination of them, inevitably there will be varying degrees of behavioral interaction between the individuals. Those interactions may be quite rigid and stereotypical, as is generally true for social insects, or they may become increasingly flexible and innovative, as is notably the case for mammals. Since even in a social group we must think of the individuals (and their genes) as struggling or competing for reproductive success, it is clear that any individual advantage will be favored by natural selection, provided it does not cause the disintegration of the society, which is in itself advantageous to each of its individual members. For this reason one can well imagine that social groups could foster a race for craft, for the cleverness of individuals so that it would raise their reproductive potential within the group. This would be a straightforward consequence of natural selection. The inevitable result would be a progression of increased intelligence and brain size, as one finds in dolphins and primates, ultimately in the hominids. It is important to remember that these societies of individuals with greater mental abilities have not outcompeted those with lesser ones; all seem to have their own ecological niche, and we find social animals of all levels co-existing on earth today. Each social level has qualities that make it a stable product of natural selection. There have been extinctions, for the world is ever changing, but the primitive societies seem to be as successful as the most complex one, that is, our own. However, who can say who the survivors will be many hundreds of millions of years from now. In our present state of world problems, I am not at all sure the ants and the termites might not have the last word. They have been successful many millions of years longer than human beings.

Awareness has allowed organisms to exert control over their life cycles. In more primitive forms it is a means of assuring or improving the chances of successful reproduction by simply responding in advantageous ways to the environment, as in slime molds finding the ideal place to fruit. This was the beginning of a trend that led to animal behavior where the control of the life cycle for reproductive success has become increasingly elaborate. It has led to the production of life cycles that involve animal societies, taking advantage of the benefits of collective action, and it has led to the production of large brains to make control over the cycle even more sophisticated within the social groups.

Chapter 8

BECOMING CULTURAL

FINALLY, we come to the ultimate achievement of the adult stage of the life cycle. It is the advent of culture. As we shall see, this involves a different kind of evolution than the one we have discussed so far.

With time, everything evolves, but the changes are not always caused by the same forces. The shape of a boulder is determined by the elements; the shape of animals and plants is changed because of natural selection; our customs and manners change in what is called cultural evolution, which is achieved in a quite different way from the evolution of shapes of either rocks or organisms. The difference between evolution by cultural selection and evolution by natural selection is sharp, and it is important to understand the distinction and keep it clearly in mind.

In evolution by natural selection there is a selection of genes, where the successful ones are perpetuated and the unsuccessful ones become extinct. Culture, on the other hand, is something quite different; it is not defined by a gene. It is in the realm of behavior and involves ideas, fads, learning, customs, traditions—all those things which can be described as behavioral information. Genes do not govern any of these facets of our lives; culture is concerned entirely with the way we behave and what we do, say, and think. Behavioral information is not inherited from one generation to the next, but is learned by one individual from another. Therefore the passing of behavioral information between individuals is what we mean by culture, and the changes of that information over time is cultural evolution. Richard Dawkins has suggested that since we call the inherited information that is passed from one individual to another in natural selection "genes," the behavioral information that is passed between individuals in cultural change could be called "memes." This is a convenient bit of short-hand, for among other things it helps to remind us of the very big difference between the two.

I should note here that my defining culture as the passing of information from one individual to another has greatly infuri-ated some anthropologists. They would prefer, in the first place, to confine the term to human beings and not dilute it with any talk of animal behavior. They think of culture as all the behaviors in any one human society. I see no great prob-lem with saying that culture involves many, diverse behaviors, but I become a bit huffy in return when I am told that what animals do bears no relation to what human beings do. This disdain of biology may have been justified in the early days of this century when cultural anthropology was in its infancy and undermined by a most unfortunate popular spasm of many bi-ologists for eugenics, the genetics and the breeding of human beings by some sort of perverse unnatural selection. As a result the more sensible anthropologists had no choice but to sever their connections with the eugenically minded biologists of the time; I like to think I would have done the same thing myself. But that was long, long ago and most biologists have become more level-headed on these sticky issues, as indeed many anthropologists now appreciate. But for numerous rea-sons the rapprochement is hardly complete, as I know from the many shafts that came in my direction when I first expressed such ideas. Furthermore, we should also remember that cul-ture is used in many other ways as well, from horticulture to what the Oxford English Dictionary refers to as a refinement of mind, tastes, and manners—a rather Victorian view of the matter.

Memes and genes have some very distinctive properties. Evolution by genes is exceedingly slow compared to evolution by memes. Genes can be passed from one individual to another only via the egg and sperm, from one generation to the next. To change the genes in a population of individuals therefore takes many generations. Memes, on the other hand, can shift to and fro with the greatest of ease and speed. For instance, a new fad, be it in dress style or a new game, can spread through a population very rapidly, and disappear with equal speed. Even faster meme transmission can occur by word of mouth in the form of a rumor. But memes need not always be rapid. A good example comes from styles of dress: the length of women's skirts may go up and down with considerable speed in most of the Western world, but among traditional Amish communities or in South Sea island societies the skirt length—indeed all the clothing—has not changed for many centuries. Yet these too can change rapidly should the society be infected by new ideas.

The other fundamental difference between the two is their physical existence. Genes are encoded in the DNA of the cells, while memes are encoded in the activities of the nervous system, which is in turn encoded by the genes during the embryonic stages of development of an individual. Put in another way, we can have organisms, such as plants, that have genes, but no memes. The reverse is impossible: there can be no such thing as an organism without genes.

. . . . .

We are now ready to begin thinking about evolution, and I will divide what I have to say into two parts. First I will discuss how the ability to have culture might have evolved, and then I will go on to the matter of how culture itself evolved.

I have already said that we are not the only species that passes on information from one individual to another; in fact, it is quite widespread in the animal kingdom, and has been most studied in the behavior of vertebrates. So if we look at different animals we can see different degrees of skill in the passing of memes between individuals.

The big question is: Why are there memes at all? Not all organisms have memes—in fact, as I just pointed out, the entire plant kingdom manages quite well without them. The answer must be that memes can do things that are impossible to achieve with genes alone, and this advantage of animal culture is easy to show, as I do presently. However, note that the ability to have memes has arisen through natural selection. The nervous system itself is a child of natural selection, and if memes are advantageous in selection, one can easily assume that those brains which were most adept in transmitting and receiving memes were favored and inherent in individuals that were successful in reproduction. Let me now give some examples where the advantage of memes is obvious and the argument compelling.

My first example involves bird migration. It is well known that many birds, such as ducks and geese, not only migrate great distances between the summer nesting grounds and the wintering grounds, but they will take the same route or flyway every year. For instance, this has been especially closely observed in snow geese by Frederick Cooke and his associates. They nest on the shore of Hudson Bay in northern Canada, and winter in southern Texas. After hatching, the young birds grow rapidly, and by the end of the summer they follow their parents south on the traditional flyway. In their first spring they again follow their parents to the home nesting ground, that is, to precisely the same region on the Hudson Bay shore. The young birds then strike out on their own and seek mates. Come the following fall (or the one after that) they are able to fly along the same routes without guidance from their elders— in fact, they now guide their own young.

We assume that this behavior is advantageous—or more likely essential—for survival. Now imagine the problem of encoding all this geographic information in the genes. Clearly the idea of having genes which can specify a large portion of the geography of North America is absurd; it would be an impossibility. Yet by memorizing the geography from one or two experiences, the birds can make the seasonal trips with the greatest accuracy. The only contribution of the genes is to produce a brain that is capable of memory and the instinct to follow one's parents when young. That geese can learn new bits of landscape (other than their ancestral ones) was shown in a delightful anecdote of Konrad Lorenz. He had a pet flock of greylag geese that he would take swimming every afternoon during the summer. One day at the appointed time he decided to desert them and crawled up on the roof to see what they would do. After a long period of honking on the front lawn, making clear their indignation at having been stood up, they took to the air and first flew over the pond where they had gone with him most frequently and then on to successive ponds where they had gone with decreasing frequency, before they gave up all thoughts of a dip with their master.

There are many examples of the use of memes to rapidly exploit new sources of food, or new ways of getting the food more easily. For instance, there is the famous case of Britain's blue tits (which are similar to our chickadees), which learned to peck open the aluminum foil on milk bottles. In Great Britain milk is not homogenized and therefore the cream settles to the top, making it easy for the clever birds, once they learned to break the seal, to reach the richest part. The outward spread of this invention was measured as new birds learned by watching their neighbors and adopting the trick for their own use. Pecking open milk-bottle caps is a most successful meme, and it later led to a human meme. People began to put out plastic cups with the empty bottles, so the milkman could cover the tops of his early morning deliveries. I was told recently that skimmed milk with blue caps is now also sold, and that the tits have already learned not to bother with blue-capped bottles. Again one can see that such fast learning of a way to feed efficiently can spread rapidly. Pasteurized milk in bottles is a quite recent invention. If the cream-eating behavior had to be achieved by the direct genetic control of behavior, it would take centuries to become established, if ever. Besides, the British milkman is likely to alter his containers some time in the near future, as he did in the United States.

Another celebrated example is that of Imo, the young female Japanese macaque that made two notable inventions. Macaque monkeys were kept on an island off the coast of Japan, and their food was dumped by boat onto the beach. Imo learned to take the sand-covered sweet potatoes and wash them in the ocean before eating. Later, in a more sophisticated undertaking, with wheat kernels, she would take the wheat into her hands, go down to the sea, and drop it into the water. The sand would fall to the bottom, but the wheat floated and she skimmed it off the surface. Her clever tricks spread through the colony, first to the younger monkeys and later to the more conservative and staid adults.

A particularly interesting example is the fishing method of green herons in ponds in Japanese parks and temple gardens. Visitors are forever throwing bits of bread or biscuits into the water to feed the fish, and it was noticed that somehow the herons had learned the same trick. They would throw in a morsel as bait, and when a fish rose to take it, the heron would, with its rapier thrust, grab the fish. Green herons use not only real food, which is fairly difficult to find, for bait, but also things such as bits of paper, leaves, twigs, fruit, plastic, and— best of all—small feathers (human beings are apparently not the only fly fishers!). Not all green herons do this, but only some of those that live near these ornamental ponds. We do not know if they learn solely from one another, or from watching people. In any event this is a splendid example of cultural or meme transmission in nonhuman animals.

In addition to providing ways of more efficient feeding, memes are also used in avoiding predators. An especially good example of this comes from studies on mobbing in birds. One frequently sees smaller birds gather around a larger predator and make the most tremendous fuss, both in noise and often in aerial acrobatics. It might be crows mobbing an owl or a hawk, or kingbirds mobbing a crow; it is a very common sight. Apparently the purpose is to draw attention to the predator so there can be no attack by stealth, and for this reason mobbing is thought to be of considerable advantage for the survival of the prey.

The phenomenon of mobbing has been studied by some German workers in an ingenious experiment. They put a partitioned cardboard box between two cages, each containing a European blackbird, which is very similar to the American robin. One of the blackbirds saw a harmless, stuffed Australian honeyeater, while the other saw a stuffed owl. The latter went into a frenzy of cries and fluttered in its cage, for owls are its mortal enemies. After a bit, the other bird that could only see the honeyeater began, by imitation, to do the same thing. Later this bird and a new blackbird were put in the opposite cages, and this time they were both presented with the Australian honeyeater. The bird that had learned to mob it immediately began to do so, and soon the untutored bird started to do the same. In fact, they repeated this experiment so that six naive birds learned, one from another, to mob this new predator (at least, that is what they were tricked into thinking). The advantage of such a cultural exchange is that the bird can quickly recognize any new predators that might appear in its environs. If the behavior which involves the recognition of a foe had to be genetically inherited, the population might be wiped out well before the necessary genes had settled into place.

This is a particularly instructive example because it is well known that it is possible to have genes which control behavior so that a particular predator is recognized. In some old experiments in animal behavior, it was shown that young goslings, just hatched, will rush under cover if a silhouette of a hawk is passed over them on a wire. But if the silhouette is turned around so that it looks more like a flying goose, the young pay no attention. In this case there is a clear advantage to having inherited the recognition behavior, for there is no time to teach the goslings the nature of this particular danger. So predator recognition can either be genetically or culturally transmitted, depending apparently on which is most useful for survival in a particular instance.

Another most interesting example of the problem comes from Darwin's discovery of sexual selection. Here there is a conflict between selection for protection from predators and selection for success in mating in some species. The result of sexual selection is that in many species there is a great difference in the physical appearance of the sexes.

While Darwin's greatest contribution is undoubtedly his deep understanding of natural selection, he also brought us the idea of sexual selection. Briefly, it means that in many species there is a great difference between the sexes. In some species the female is larger than the male, and in others it is the reverse. In the best-known cases the male is beautifully adorned, as in the peacock, where the peahen is comparatively drab. Darwin argued that the main reason for this difference (or dimorphism) must be that the female chooses to mate with the most brightly colored male and ultimately this produces the excesses seen in birds of paradise or peacocks. (There are other instances where sexual selection may be propelled by the competition between males, but this more often leads to larger and stronger males rather than more ornate ones.)

There have been many tests to show that fancy male birds have an advantage over the more unassuming ones. A good example has been provided by Malte Andersson on the African widowbird. The males have an absurdly long tail, and Andersson cut off part of the tails of some birds and glued them on to the tails of others to make them abnormally long. He was easily able to gauge the reproductive success of these birds, because the males defend territories, and the larger the territories, the more females and nests they acquire. He found that the birds with the artificially elongated tails had the most nests in their large domains, the birds with clipped tails had the fewest nests, and the normal birds were intermediate in their success. I cannot imagine a neater and clearer demonstration of sexual selection, all based on the visual cues sent from the males to the females.

There is another example of this phenomenon that has always fascinated me. Birds of paradise are perhaps the most extreme example of this kind of sexual selection. The males have an incredibly elaborate plumage, with which they puff out and posture to produce the most astounding displays. In numerous species the males will gather together in one tree and go through their elaborate strutting together, and the females will come and pick the best performer. In all these cases it is assumed that it is not just the male splendor that is selected, but that somehow splendor is associated with other correlated traits that represent vigor in some form. The female wants to mate with a male that will sire offspring that have the greatest chance of survival in subsequent generations—as one would expect of selfish genes.

Bower birds are directly descended from birds of paradise, but they are quite drab in appearance. An interesting suggestion was made by Thomas Gilliard that they make up for their dull plumage by building elaborate bowers, which can be quite extraordinary. The simplest ones are avenues of sticks, but the more complicated ones, the maypole bowers, are an impressive tent of sticks surrounding the base of a small sapling. Besides having been given an extraordinarily intricate structure, the bowers will be adorned on the ground with colorful objects at their entrance. In early days in New Guinea and Australia, this festooning consisted of flowers, snail shells, bones, fruit, and bright leaves, but with the advance of civilization the bowers have become adorned with such objects as keys, bottle tops, toothbrushes, false teeth, spectacles, buttons, brass cartridge cases, teaspoons, coins, broken glass, nails, Kodak film boxes, and any other colorful artifact of modern man that the birds can find. Even more surprising, some species paint the inside of the bower with the juice of berries by squeezing a berry in their beak and using it as a paint brush. (There is even an instance observed near civilization where they used blue laundry soap.) In this way they can produce blue, black, or green walls. It used to be that the Victorians were castigated for calling these structures bowers, implying that the birds had an aesthetic sense that allowed them to appreciate a beautiful love bower. Today we are not so sure they were wrong; we have regained either the courage or the recklessness of the nineteenth-century zoologists. In fact, Jared Diamond, using poker chips, has recently shown that different groups of the same species in different valleys in New Guinea had their own preferences for color: one group would prefer blue, another red. He concluded that this preference was cultural and was passed on by imitation within a population, which suggested to him some sort of rudimentary aesthetic sense.

The building of the bower itself seems to involve slow learning and much trial and error by the males. When an experienced bird finally finishes building a bower, the courtship itself becomes of great interest. Let me quote a passage from an article of Jared Diamond's which describes the process admirably.

> Bowers have traditionally been interpreted in the former sense, as love parlours where males seduce females. Vellenga's observations provide ample evidence for this function. The bower-owning male Satin Bowerbird continually tries to entice females into his bower by picking up in his bill an object such as a flower or snail shell, and posturing, dancing and displaying to the female. Rarely, the female is persuaded to enter the bower and copulation ensues. Far more frequently, the female flees without mating.
>
> Why do courting males have such a low success rate? Perhaps females scrutinize various males and bowers before choosing. Many courtships are interrupted at a crucial stage by the intrusion of other individuals. However, another reason for the low success rate may be the invariant sequel to successful mating: the male bowerbird savagely attacks the female, pecks and claws her, and chases her from the bower. Mating itself is so violent that often the bower is partly wrecked, and the exhausted female can scarcely crawl away. The courtship display can appear little different from the male's aggressive display. When a courted female is won over and starts to solicit copulation, the male often changes his mind and chases her away. Thus, a female may have to make many visits to a bower before she overcomes her fear of the aggressive male. After mating, the female constructs a nest at least 200 yards from the bower and bears sole responsibility for feeding the young.

To return to Gilliard's interesting idea, he showed that there was an inverse correlation between the degree of coloration of the male of a particular species and the elaborateness of his bower. In other words, the ancestral species, like their relatives the birds of paradise, had a colorful plumage and a simple bower. As the bower became more complex, the plumage of the males became increasingly drab and differed little from that of the female. Note, however, that the male can still attract the female by a gaudy display—not with its own body but with its bower. Recent work has shown that experienced birds with the best bowers attract more females who mate with them than those with inferior accommodations. In fact, these dominant birds will sometimes go on forays and utterly wreck the neighboring bowers. A further advantage of this transfer of sexual feather finery into architectural behavior: they are no longer so conspicuous and at the mercy of predators.

The important point is that sexual selection often seems to be working in a direction opposite from that of natural selection. Sexual selection leads to increasingly conspicuous males in the examples I have given, and while this may make them successful in mating, it increases their chances of being discovered and devoured by a predator. Seen this way, it is clear that the bower birds have devised a very clever way of avoiding a conflict between these two kinds of selection. It is a way of accommodating sexual and natural selection simultaneously.

. . . . .

Now we can discuss an even more general point about behavior and natural selection. It would seem reasonable that if an animal behaves the same way (through selection) for numerous generations, genes could seep in by chance to fix those behavioral traits. Such a process, as I said earlier, is called the "Baldwin effect," and it seems to be a very common property of genes. If there is selection for a trait, or no selection either for or against a trait that keeps repeating itself in successive life cycles, then there are good chances that genes will appear that ensure the same thing. There may be no need to have two ways of establishing a trait, but genes will accidentally filter in under such circumstances simply because there is no selection against them. We have already described a similar process for development. It is even quite possible that the difference between regulative and mosaic development could be explained by genes fixing the developmental steps that lead to mosaic development. But then, under some circumstances it must be advantageous to remain regulative, as for instance in regeneration, and then selection favors the removal or prevention of the rigid genetic fixation of the developmental steps. Similarly, in behavior it is advantageous, from the point of view of natural selection, for birdsong to be learned, as happens in many species. In others, such as the cuckoo and the cowbird, there is strong selection pressure for the song to be genetically determined, otherwise no mating would be possible. My personal and somewhat unorthodox view is that this kind of seeping gene fixation is a far more important and widespread process than is generally recognized.

．． ． ． ．

A final example of behavior that is connected with the threat of predators is tameness in some wild populations. Human beings are among the most formidable predators of all wildlife, but in some places on earth they have long been absent. The Galapagos Islands are one such spot, and Darwin, when he visited there in the early part of the nineteenth century, was greatly struck by their remoteness. In his *Voyage of the Beagle* he wrote: "Formerly the birds appear to have been even tamer than at present. Cowley (in the year 1684) says that the 'Turtle-doves are so tame, that they would often alight upon our hats and arms, so as that we could take them alive: they not fearing man, until such time as some of our company did fire at them, whereby they were rendered more shy.'"

There is considerable evidence that shyness can be learned, and, for that matter, so can a reversion to tameness. It is well known among farmers that crows will keep a greater distance between themselves and a man carrying a shotgun than one who is walking empty-handed. In bird feeding stations or in parks, patient people can gradually teach birds to take food from their hands; the absence of danger or threat of any sort allows the bird's appetite to overpower its fear. One of the best examples of shyness becoming a tradition is given by A. and O. Douglas-Hamilton in their book *Among the Elephants:*

> Another example of their traditional learning comes from the South African National Park of Addo. Here, in 1919, at the request of neighboring citrus farmers, an attempt was made to annihilate a small population of about 140 elephants. A well-known hunter named Pretorius was given the job. Unlike Ian Parker's teams, whose rapid semi-automatic fire liquidated entire family units, Pretorius killed elephants one by one. Each time survivors remained

who at the sound of a shot had witnessed one of their family unit members collapsing dead or in its death agonies.

No doubt the trauma was often associated with the hunter's smell, and danger arising from the presence of man had become deeply imprinted in the memory of those who escaped. Within the space of a year there were only sixteen to thirty animals left alive. It seemed that one final push would rid the farmers of their enemies, but by then the remaining elephants had become extremely wary and never came out of the thickest bush until after dark. On several occasions when the hunter ventured to pursue them, he in turn was hunted through the dense thickets and had to flee for his life. Pretorius eventually admitted himself beaten and in 1930 the Addo elephants were granted a sanctuary of some 8,000 acres of scrubby hillside.

The behavior of these survivors has changed very little, though they have been contained by a fence and are not shot any more. Even today they remain mainly nocturnal and respond extremely aggressively to any human presence. They are reputed to be among the most dangerous elephants in Africa. Few if any of those shot at in 1919 can still be alive, so it seems that their defensive behavior has been transmitted to their offspring, now adult, and even to calves of the third and fourth generations, not one of which has itself suffered attack from man.

It is clear that tameness is another example of the Baldwin effect, for some populations, such as many of the birds and other animals in the Galapagos Islands, seem to remain tame despite the threat of man. Perhaps they have been without threat for so long that tameness genes have seeped in and govern their behavior. Man's arrival in the islands has been so recent in terms of evolution that there has not been enough time for those genes to be selected out.

• • • • •

I would now like to give some attention to human beings: what is so special about us? I hope I have made a good case that nonhuman animals can pass on behavioral information from one individual to another, and in this sense they have culture. In my view, the big difference between them and ourselves (and it is a huge difference) is one of quantity. We are so much better at culture transmission that the gap between us and even the great apes seems enormous.

I will now explain the more obvious of those differences, and then go on to show that because of them we have been able to have an evolution of culture, something that can hardly be said of other animals. Nonhumans might go from one tradition to another, and in this way make a minor step towards a cultural evolution, but in the case of humans it is quite different. We have a cultural history that, since the early beginnings of our species, has changed and evolved into the way we live and think today. It is a gigantic evolution and has happened in a very short space of time (for it involves rapid memes and does not depend on gene-limited natural selection). Our cultural evolution is not the result of any new property that we possess, but we are simply better at passing memes, and the improvements we have made in such meme handling (that is, bigger brains that arose by natural selection) have led to more memes—so many that they have virtually taken over our lives and our destiny. If one were to argue that we are indeed different from other animals, and among other reasons that we have a history, an evolution of our culture, I can only agree. However, I would insist that the only reason for this is that we have evolved by natural selection into an animal supremely efficient at passing and accumulating memes. The ways in which we do this are fundamentally no different from that of other animals; we have become different by doing more of the same thing.

The most obviously useful skill we possess is language. Language can be used easily to pass information from one individual to another: it is the ultimate meme transmitter. This aspect of language is certainly no different from what we find in animals. To give an example, it is well known that alarm calls can sometimes be quite specific as to the danger. For instance, vervet monkeys have three distinct alarm notes, which was proven by recording their alarm calls and then playing them back to a group of monkeys. One call clearly means "hawk," for when it issues forth from the loudspeaker all the vervets look up into the sky. After another distinct call that means "leopard," the monkeys run out in the open glancing at the neighboring thickets. A third call meaning "snake" causes the monkeys to stand on their tip-toes, searching the ground around them. These are only three "words," and no doubt vervet monkeys have several more. There also have been observations of chimpanzees and gorillas in the wild who emit all sorts of grunts and muffled noises. It is presumed they impart information, but mostly we have not yet cracked their code. Based on the number of sounds they make in the wild, the estimate for chimpanzees is in the order of thirty signals or "words." But even if there were two hundred, they would literally be a far cry from the vast vocabulary of our own species.

Everyone agrees that our language skills differ from animals in the number of signal words. We do have a larynx and vocal cords that are particularly suited to speech, something that great apes lack. It is, of course, true that parrots and numerous other birds are splendid mimics of human voices, and there are examples where they knowingly use our words as signals; but again, the number any one parrot can learn is comparatively minuscule. Also, chimpanzees and gorillas have been taught to communicate by American sign language (for the totally deaf) and by other nonvocal methods, but again they too learn a vocabulary of not much more than about two hundred words.

However, it is not just that our vocabularies are richer; we can also put sentences together. There is a great debate as to whether trained chimps have this capacity to achieve some sort of simple syntax, but it is a difficult, technical argument which I do not fully understand myself. In any event, we do not know yet what great apes can achieve as grammarians and whether syntax itself has its origins in animal behavior, only to flower in human beings. If this were so, it would clearly suit me and others who wish to argue that we do the same things as animals but, like Annie Oakley, "only better."

Implicit in this discussion is the notion of memory: we know more signals than other animals partly because we have a better memory. However, before we end up being too pleased with ourselves, we would do well to remember that the memory capabilities of other animals are far more impressive than we thought only a few years ago. Even insects—as, for example, bees—have a remarkable ability to memorize landscape, and they can go from one part of the map to another even after they have been placed by experiment in a distant spot where they do not usually go. Some species of birds will, like squirrels, hide seeds and keep track of an amazing number of caches. The capacity to remember these hidden treasures is no less remarkable in squirrels.

I could go on with more such examples, but they all pale beside human memory. One need only think of all the facts we have accumulated by the time we graduate from high school: not only what we learned in class, but also what we learned from our fellow students (some of which may be of doubtful value). Even greater are the number of things we learned from our parents, some more willingly than others. For the first seventeen or so years of our life we are, without stop, crammed full of information, much of which we remember for the rest of our lives. But we do not stop learning; we continue to do so right into old age. Think of an actor who learns an entire Shakespearean play, or at least one of its main parts. In some people the ability to memorize is extraordinary. For example, a chemistry professor at my university awed us, to say the least, with his knowledge of the entire five-place logarithm table by heart. However, one does not have to be a memory athlete to make my point; the most ordinary, forgetful people still retain a fantastic amount of information in their brains.

Our large brain, skilled in memorizing, is only the beginning of the human success story. Because of our social existence and our inventiveness, we have managed to store information in many other ways.

In some aboriginal societies today, there is a division of labor, as there probably was in the first societies of our species. Sometimes it is divided between the sexes: the women gather and cook, while the men hunt. This arrangement can become more specialized: some will become especially skilled and knowledgeable in, say, building canoes, others in making bows and arrows, and still others in the tricks of healing wounds or chasing sickness out of the body. In more advanced and civilized social groups, we continue to have such a division of labor, and it becomes even more specialized: the cobbler makes shoes for us, the farrier does the same for horses, the tailor makes suits and dresses, the butcher cuts meat, the fisherman snares the fish from the waters, the banker hopefully keeps our money safe, the builder builds our houses, the lawyer is there to steer us through the intricacies of the law, and so forth. For each of these skills, and many others, individuals learn certain things and can serve the whole community as a repository of those skills. Not everyone has to learn every trade or occupation, but by sharing services the total memory of skills for the whole community is very much larger than what could be encompassed in the brain of one person.

There is another interesting point about human beings: unlike in most if not all other species, old people remain alive and part of the community well after the age when they reproduce. One result is that the aged have become a repository of the lore of the tribe, its customs, its traditional tales, and even of survival information that might be useful in times of crisis. In this way they serve as special memory banks for the benefit of the social group. They are the voices of experience, the carriers of tradition and history, and therefore the voices of wisdom. As one matures, this idea has increasing appeal.

There is considerable debate as to when early man invented speech and language. But all of it is conjecture because, although fossil remains are becoming increasingly abundant through many remarkable anthropological discoveries, we do not have what is needed: a good series of fossil larynxes extending from our primate ancestors to modern man.

What we do have, however, is a record of some of the early ways of congealing language in the form of writing, and here is an invention of major importance for the storage of memory. It is now possible to put a fact or an idea into the form of a symbol, and we only need to remember the symbols, the letters and the words. Then at any time later we can look at the writing and revive the stored memory. A student in a lecture course takes notes so that she has a record of the golden words of the professor. This is a great boon, for everything does not have to be memorized the moment it is spoken and heard. The student can review her notes before the examination and (if her handwriting is better than mine) achieve total recall for the final test.

The first evidence we have of writing is from Assyrian cuneiform, a notation which involved the imprinting of symbols (letters and words) in soft clay that became permanent upon hardening. Unfortunately not all of the tablets that have been discovered in the archaeological digs of this ancient civilization reveal facts of cosmic importance, for many are simply accounts and inventories—sort of Babylonian laundry lists. But others are fascinating fragments about their religion and show the early origins of the Bible we use today. These are bits of information, stored for thousands of years, that we can recover and make use of today.

Devising an alphabet and imprinting it onto wet clay were two major inventions in the storage of information. The symbols for the words kept changing, or were reinvented, as in the hieroglyphs of the Egyptians or Mayans, or in the Phoenician and later alphabets which show direct and easily traceable connections to our own writing.

The process of storing information was helped greatly by various mechanical inventions. For instance, invention of the material which retains the writing made spectacular changes, from wet clay (and chiseled stone) to papyrus to parchment to the invention of paper in the Middle Ages. Consider also the evolution of writing instruments, another spectacular series of inventions, each with enormous consequences—from the stylus for wet clay and the stone chisel, to ink on the quill of a feather, to the pencil and the steel pen, the fountain pen, and now the ball point pen and all its progeny. The nineteenth century saw the birth of the typewriter, which in the last few years has become obsolete and almost totally replaced by the computerized word processor, with all its capabilities of correcting errors and even spelling. (A student of mine once handed me the rough draft of her thesis with a note attached: "Please don't pay any attention to the spelling—I haven't pressed the 'spell check' button yet.")

In addition to the better ways of writing, with the Gutenberg Bible in the fifteenth century (the first printed book) the way of copying, and therefore spreading, the written word was suddenly opened up. It was no longer necessary for scribes to copy everything over and over by hand—with movable type, a page could be duplicated exactly many times. There have been many revolutions in printing methods since then. In the past few years we have seen the death of the old way of setting type with hot lead (linotype) and its replacement by word processors, and in the photographic methods of producing the template for printing. The printing process itself has made equal advances, and a modern press can turn out copies with amazing speed. And let us not forget the Xerox copier and the Fax machine.

Since early times, books and manuscripts have been deposited in libraries which have become increasingly gigantic storehouses of information. One need only think of the tribal knowledge of some elder in a primitive society at one end of the spectrum and the contents of the Library of Congress or the British Museum on the other. The vastness of the difference is hard to grasp.

But that is not all. There is another series of modern information storage devices of even greater power. It began with photography, and then came the recording of moving pictures on film, which is the history of Hollywood. The first ones were without sound, but then they became "talkies." The early films were mostly playful fiction and often achieved greatness, as with the movies of Charlie Chaplin and Buster Keaton. But some also recorded current events, and these are now of great historical importance: we can see horse-drawn streetcars; the slums as well as the beautiful parts of our cities, with people walking around in the clothes of the time; and we have extraordinary records of events such as the First World War. The old films and cameras (and the lenses of those cameras) were constantly improved, including the advent of color film and processing. The latest transmutation has come in the magic of video, where one can take a color moving picture and immediately see the result on a television screen.

Finally, there is the computer, which, as an information-storing device, beats everything, not only because of the amount it can store, but because of the ease of retrieving information. Today you can call up the bank and give your account number. Immediately the banker has your account on the screen with all the latest deposits and withdrawals and, just as you feared, you are overdrawn. Or, to give a biological example, now that it is possible to sequence genes down to the detail of their series of bases, you can put whole gene sequences (which are very long) in a data bank. Then, if you discover a new gene and find its sequence of bases, you can plug it into the data bank and ask what genes it resembles among all the thousands of genes already stored there. In this way some remarkable discoveries have been made about the similarity of some genes (which therefore are presumably of common origin) that had hitherto been totally unsuspected. This has resulted in important advances in molecular biology in the last few years. To end this paragraph with a somber note: all our income tax returns of the past few years are stored on computer tape and are at the beck and call of the Internal Revenue Service. So, although our ability to store information is great and our collective, machine memory is beyond our wildest imagination, we cannot always be confident that the most interesting information is being stored.

Some attempts have been made to record the total progress humans have made during the course of civilization. Clearly the accumulation of information was relatively modest in our early history, but it has increased at an alarming speed. From the data of two studies on the rate of making inventions between A.D. 1300 and 1900, it is clear that it rises exponentially at a rate of 25 percent per century. Compare this with the rise in population during the same period: it rises at only 2.4 percent per century. One wonders if the time will come when we have so much information that we do not know what to do with it all. I suspect that as long as we stay in control and don't let the computers take over, we will manage to edit and keep what is useful and bury the rest.

.　.　.　.　.

Now that we are at the end of this journey, let me summarize in a grand manner, but with few words, what I have said and done here. This book is about the life cycle, and yet it is also a sort of intellectual or biological autobiography. To put it another way, a large element in this book is my own life cycle.

It has become my conviction that the most fundamental and interesting aspect of all life is the life cycle. It was conceived by the ability of the genes to duplicate themselves by virtue of the properties of DNA and natural selection. Once a cycle was established it could vary because of the mutable properties of DNA which ultimately, through natural selection, became stabilized in the form of sexuality. Then arose the possibility that more than one kind of life cycle exists on earth and that each could successfully exploit a portion of the environment without necessarily causing the extinction of the other. There was a continuous selection pressure during the last few billion years for increasing the variety, which means increasing the complexity and size of life cycles, so that we find an array of millions of animals and plants of all sizes and shapes inhabiting the earth today.

The life cycle is not the only cycle. There are cell cycles within a multicellular organism, and there are cycles of multicellular societies. However, these and other less obvious cycles are all subservient to the life cycle, for the latter, through the genes, governs all the lesser cycles within an organism and all the greater cycles, such as those of animal societies. Finally, it was by building on the foundation of the life cycle that behavior in animals became possible, again by natural selection, and behavior led ultimately to cultural evolution, which has so greatly influenced our own lives today.

SELECTED READING

SINCE this book is meant to be read but is not a reference book, a textbook, or a book which takes kindly to footnotes, here, in a short essay, I will make some suggestions for further reading on some of the many subjects touched on in the preceding pages.

There are many splendid books on the history of biology, and in particular on the history of the study of evolution. For instance, there are large numbers of books on Charles Darwin, many of them excellent. One of the most recent products of what is now called the "Darwin industry" is a riveting book called *Darwin* by Adrian Desmond and James Moore (Michael Joseph, 1991). Do not be put off by its large size—it will be good for your stomach muscles should you read it in bed. It not only traces the life of Darwin, but at each stage of his career it gives you a vivid picture of all sorts of other things happening at that time, such as the politics of Britain and the rest of the world, the biology of the day, and the emerging politics of science. There is also a fascinating description of the struggles between the Anglican church and evolutionary biologists. All this is done with great insight into the mind of Darwin himself. The reviews of this book have been excellent, although one historian of science complained that it had everything except information about Darwin's sex life!

For those who want to know more about Darwin's great voyage, his *Voyage of the Beagle* (Dutton; Natural History Museum; Bantam) is highly recommended. It is a splendid book. Equally fascinating, however, and perhaps an even better book, is Alfred Russel Wallace's *The Malay Archipelago* (Dover). I would be remiss not to mention also Henry Walter Bates's *The Naturalist on the River Amazons* (Dover). The nineteenth century was the era of the literate and intrepid explorer.

It is far more difficult to find easy books to read on the subject of developmental biology, for today it is such a technical and rapidly moving field, especially with the advent of molecular biology. If you want to get a better feel for what I am saying, go to the library and look at *The Molecular Biology of the Cell* by Bruce Alberts et al. (2d ed., Garland Publishing). It is not a book one reads any more than the *Encyclopaedia Britannica*, but it will give you a peek at modern experimental biology.

The subject of evolution, on the other hand, has a rich supply of excellent books. Despite its importance, Darwin's *On the Origin of Species* is not a good place to start, although I can highly recommend it for eventual reading. I suggest reading the first edition of 1859 (facsimile ed., Harvard University Press), which is far simpler and easier to read than any of the subsequent ones (six in all), in which he tried to satisfy his contemporary critics. A better place to begin is perhaps with George Gaylord Simpson's *The Meaning of Evolution* (Yale University Press, 1967), but I do not find the philosophical portion at the end rewarding, unlike the rest of the book. Two clear, but more difficult books are John Maynard Smith's *The Theory of Evolution* (Penguin, 1958) and Philip M. Sheppard's *Natural Selection and Heredity* (3d ed., Hutchinson, 1967). Especially worthy of mention is Ernst Mayr's book, which is available in two forms: the original, complete book, *Animal Species and Evolution* (Harvard University Press, 1963), and a more popular version, *Populations, Species, and Evolution* (Harvard University Press, 1970). Mayr writes clearly and somehow manages to convey the grandeur of evolution.

In the domain of behavior there is again a vast array of splendid books—it is hard to know where to begin. For sheer pleasure, with wonderful insights, Konrad Lorenz's *King Solomon's Ring* (Crowell, 1952) cannot be too highly recommended. It is a gripping introduction to modern animal behavior (ethology) by one of its founders. Niko Tinbergen, another founder, wrote *The Study of Instinct* (Oxford, 1951), which is in its own way an even more important book. It is a systematic discussion of the beginning of ethology and is written with such clarity that it shines. Karl von Frisch wrote up his magic observations and experiments on bee behavior in *The*

Dancing Bees (Methuen, 1954). There are many more recent books including some excellent textbooks. To mention one, James Gould's *Ethology* (Norton, 1982) gives a clear and authoritative review of where we are today. Of the many studies on primates I would like to make two suggestions: Jane Goodall's *In the Shadow of Man* (Houghton Mifflin, 1983), and Franz de Waal's *Chimpanzee Politics* (Harper and Row, 1982). If one is interested in what goes on in the mind of an animal, I recommend Donald R. Griffin's *Animal Thinking* (Harvard University Press, 1984), a book full of interesting insights.

On social animals, nothing beats Edward O. Wilson's *Sociobiology* (Harvard University Press, 1975). It has the unique quality of being both encyclopedic and readable. Do not be alarmed by its size (although it would be unwise to try reading it in bed). I can also highly recommend an exceptionally well-written book, full of riches—Robert Trivers's *Social Evolution* (Benjamin-Cummings, 1985).

Finally, in this book I frequently discuss matters which I have considered in much greater detail elsewhere. I will list the sources here, in case the reader wants to pursue any of these matters further and needs references to the literature. I do this with some hesitation because of a point my colleague Henry Horn has made about citing oneself. He says that one should not assess a person's productivity solely from the Citation Index (which lists all the places where any paper or book is cited), that is, by taking the total number of citations for a particular author. Instead, one should divide the number of a person's citations by the number of times he cites himself! In what follows I will be giving myself a rather low score. I have discussed life cycle matters in *The Evolution of Development* (Cambridge University Press, 1958), in *Size and Cycle* (Princeton University Press, 1965), and in *The Evolution of Complexity* (Princeton University Press, 1988). Problems of size were examined superficially in *The Scale of Nature* (Harper & Row, 1969), and more seriously in T. A. McMahon and Bonner, *On Size and Life* (Scientific American Books, 1983). I have discussed problems of development in *Morphogenesis* (Princeton University Press, 1952) and in *On Development* (Harvard University Press, 1974). In *The Evolution of Culture in Animals* I concentrated on behavior and the beginning of cultural evolution. Finally, I gave a general survey of the earlier work on slime molds in *The Cellular Slime Molds* (2d ed., Princeton University Press, 1967).

INDEX